Sky & Telescope's

POCKET
SKY ATLAS

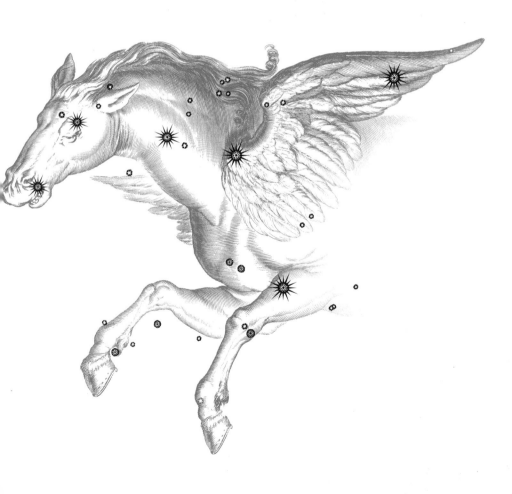

Sky & Telescope's

POCKET
SKY ATLAS

Roger W. Sinnott

Sky Publishing Corporation
Cambridge, Massachusetts, USA

Library of Congress Cataloging-in-Publication Data

Sinnott, Roger W.
  Sky & Telescope's pocket sky atlas / Roger W. Sinnott.
    p.   cm.
  ISBN 1-931559-31-7 (alk. paper)
  1. Stars — Atlases. 2. Astronomy — Charts, diagrams, etc. I. Title: Sky and Telescope's pocket sky atlas. II. Title.

G1000.2S5 2006
523.8022'3 — dc22
                    2006042290

Contents

Introduction

Think of this as your *working* sky atlas, the one you carry out to your backyard telescope or pack in luggage for a trip far from home. Today, amateur astronomers have abundant choices in the celestial charts they use. Some come in elegant multi-volume sets, a few have large sheets that unfold, and others run on a computer. But did you ever see a booklet of star maps so handy to use at the telescope?

The editors of *Sky & Telescope* magazine don't just write. Like amateurs everywhere, we go out at night to observe stars, planets, the occasional comet, and a profusion of deep-sky sights. In a half century we've published six major astronomical atlases, but this — our seventh — is the only one designed and prepared entirely in-house. We've paid close attention to features our readers have begged for in an atlas over the years.

Above all, we wanted this *Pocket Sky Atlas* to be convenient for use in the field at night. Charts on facing pages form a spread that spans 50° of sky, but you can also fold the book back on its spiral binding and hold it in one hand. For recognizing star patterns quickly, we've added "stick figures" for constellations — the same ones we've used in *S&T* since 1993. These green lines should help you find your way better than the official constellation boundaries (finely dotted lines).

The Milky Way is portrayed by a simple two-tone scheme in which a darker shade of blue marks the brighter core. These contours are adapted from the first (1981) edition of Wil Tirion's *Sky Atlas 2000.0*.

What the Atlas Shows

The 80 main charts contain 30,796 stars to visual magnitude 7.6, roughly five times as many as you can normally see with the naked eye. Because all these stars are readily visible in a low-power finderscope, they make helpful benchmarks and patterns for "star-hopping" to any object of interest.

Double and multiple stars are marked with a horizontal bar. For the purposes of this atlas, stars are considered double if they have a companion brighter than magnitude 10.5 that is resolvable in amateur telescopes within 30″ of the primary star, or a companion brighter than 8.5 within 200″.

Variable stars are shown with an outer ring to represent the star's brightness at maximum, and an inner dot to show the star's brightness at minimum (if that minimum is magnitude 7.6 or brighter). Only those stars that vary more than 0.5 magnitude are identified as variables here.

The selection criteria for nonstellar objects depend on the type. Virtually all the galaxies plotted are brighter than visual magnitude 11.5 and globular clusters brighter than 10.5. Planetary nebulae, whose light is typically more concentrated and hence easier to see, are included if brighter than about 12. The light of an open star cluster tends to be spread over a generous patch of sky. Nearly every cluster we've included has an overall brightness exceeding that of an 8th-magnitude star, but a few dozen fainter clusters are included because they are rich or compact and just as easy to see.

The outlines of most bright nebulae, supernova remnants, and dark nebulae have been adapted and simplified from those drafted by E. Talmadge Mentall for the *Millennium Star Atlas.* Some of these objects, like the magnificent Simeis 147 at the Taurus-Auriga border, are difficult or impossible to see visually but rewarding to capture photographically.

We decided to include all objects in the Astronomical League's popular Herschel 400 observing list, even though a few of them would have been rejected by the selection criteria outlined above. For a special observing treat, try hunting down some of the 55 carbon stars identified by "(c)" in the atlas. Their vivid red hue can be quite striking in a large amateur telescope.

For more information about deep-sky objects of all kinds, and astronomy in general, consult the books and links in the Bibliography on page xi.

Understanding Labels

Many bright stars have a traditional name like Betelgeuse or Albireo. A larger number carry a Greek letter assigned by German cartographer Johann Bayer in 1603, or perhaps a roman letter (especially in the southern sky). For important stars not identified this way, we use the number applied by English astronomer John Flamsteed early in the 18th century (though many Flamsteed numbers for stars fainter than magnitude 5.5 are omitted to avoid clutter). Some special-interest objects have an astronomer's name and number attached, such as the fast-moving stars Groombridge 1830 and Lalande 21185.

Variable stars are marked with an upper-case letter (R through Z), two letters (such as UV or AD), or V followed by a number. The best-known double stars have their own special nomenclature, in which the prefix indicates the discoverer. The most common of these are Σ, for Wilhelm Struve; OΣ, Otto Struve; h, John Herschel; Δ, James Dunlop; and β, S. W. Burnham.

Throughout the atlas, *nonstellar* objects are labeled in *italics.* Most nonstellar objects are known by more than one designation, but just one is given in this atlas. The 109 objects in Charles Messier's famous deep-sky list have the prefix M, and this designation takes precedence over all others. Otherwise, by far the most common numbers are those from J. L. E. Dreyer's *New General Catalogue* (NGC) of 1888 or *Index Catalogue* (IC) of 1895 and 1908. We omit the prefix NGC throughout.

When an object has no M, NGC, or IC number, some other designation is given. For open star clusters, a frequent prefix is Cr (Collinder), Tr (Trumpler), or Mel (Melotte). Many bright nebulae have the prefix Ced (Cederblad), vdB (van den Bergh), or Sh2 (Sharpless). Dark nebulae often have an LDN (Lynds) or B (Barnard) number. In the case of planetary nebulae, PK means Perek-Kohoutek. While numbers from Patrick Moore's popular Caldwell catalog do not appear in the atlas itself, their equivalent designations and chart numbers can be found in the special index on page xxiii.

Chart Arrangement

A novel feature of the atlas is its layout, illustrated on the inside back cover. Imagine the sky divided into eight gores, or lunes, running from the north celestial pole to the south. Each gore makes up a section of this book and covers a wide swath of sky passing directly overhead at a specific season of the year and time of night — as summarized on the Contents page and on the first page of that section. Therefore, the charts you'll find most useful on a given night tend to be close together in the atlas.

Within each section, chart numbers increase from north to south. From section to section, they increase from west to east. There are 10 charts per section. Notice that each north-circumpolar chart has a number ending in 1, and each south-circumpolar chart has a number ending in 0. Any chart that ends in 4, 5, 6, or 7 contains part of the celestial equator. Each chart has blue arrows in the margins, telling you the chart numbers of immediately adjacent regions.

All charts include some of the sky shown on adjoining charts, and between sections this overlap is especially generous in the polar regions. If an object lies near the outer margin on one chart, chances are excellent it will be well inside the margin on some other chart. Check the General Index on page xiii to find all the charts on which a named object appears.

Notice that the Big Dipper (Chart 32), Gemini (Chart 25), and the Great Square of Pegasus (Chart 74) are shown in their entirety. When you see how many other much-loved regions of the sky are just as well shown, you might wonder, "How could they be so lucky?" That's another secret of our chart layout! Rather than slice up the sky according to a rigid, mathematical plan, as in other atlases, we freely shifted the centers of individual charts a bit if it would help to keep important star patterns whole (while maintaining sufficient overlap between charts).

Some popular but very congested regions of the sky are shown in more detail on four close-up charts following Chart 80. These are the Pleiades, Orion's Sword, the center of what has been called the Realm of Galaxies in Virgo, and the Large Magellanic Cloud.

How This Atlas Came to Be

In the spring of 2000, fresh back at *S&T* from the Texas Star Party, Gary Seronik got the ball rolling by proposing a small-format version of Sky Publishing's best-selling *Sky Atlas 2000.0*. Within a few months he had generated test plots and convinced himself that, despite early doubts, we wouldn't have to give up a lot of faint stars and deep-sky objects, nor carve up the sky into chunks so small that the atlas would lose its usefulness in the field.

Independently, Joshua Roth was showing other editors his "dissected" copy of *Sky Atlas 2000.0*. As we kicked ideas around, I recalled my earliest days on the staff and a half-serious comment by then-editor Joseph Ashbrook. Referring to a popular book-format atlas of the time, he said it would benefit greatly by having its numerous pages of printed tables and reference material excised, leaving just the handy set of charts! Soon Richard Tresch Fienberg, Tony

Flanders, and Alan M. MacRobert — experienced observers all — joined our brainstorming sessions. Sky Publishing's Books & Products team of Paul Deans, Benjamin F. Jackson, and Kerri A. Williams launched the project in earnest with the enthusiastic support of president/publisher Susan B. Lit.

I went to work tailoring the magazine's star-charting software to the needs of this project and refining the selection criteria for stars and deep-sky objects. As soon as I could generate the actual computer plots, Gregg Dinderman brought cartographer Martin Gamache of Alpine Mapping Guild in on the project, and together they crafted the beautifully finished charts you see here. With valued input from Sandra Salamony, Gregg also designed the book, inside and out. Derek W. Corson, Sally MacGillivray, and Dominic Taormina handled countless administrative and production details and coordinated the printing by DS Graphics of Lowell, Massachusetts.

If you have any kind of telescope, you *must* have a first-rate sky atlas or you will never get very far in astronomy. Other fine atlases exist, but we think this one may come closest to meeting your night-by-night needs.

Roger W. Sinnott

Bibliography

Burnham, R., Jr., *Burnham's Celestial Handbook*, 3 vols., New York, 1978: Dover Publications.

Delporte, E., *Délimitation Scientifique des Constellations*, Cambridge, UK, 1930: Cambridge University Press.

French, S., *Celestial Sampler*, Cambridge, MA, 2005: Sky Publishing Corp.

Harrington, P. S., *The Deep Sky: An Introduction*, Cambridge, MA, 1997: Sky Publishing Corp.

Hirshfeld, A., R. W. Sinnott, and F. Ochsenbein, *Sky Catalogue 2000.0. Volume 1: Stars to Magnitude 8.0*, 2nd edition, Cambridge, MA, 1991: Sky Publishing Corp. and Cambridge University Press.

Hirshfeld, A., and R. W. Sinnott, eds., *Sky Catalogue 2000.0, Volume 2: Double Stars, Variable Stars and Nonstellar Objects*, Cambridge, MA, 1985: Sky Publishing Corp. and Cambridge University Press.

Houston, W. S., *Deep-Sky Wonders*, Cambridge, MA, 1999: Sky Publishing Corp.

O'Meara, S. J., *The Caldwell Objects*, Cambridge, MA, 2002: Sky Publishing Corp. and Cambridge University Press.

O'Meara, S. J., *The Messier Objects*, Cambridge, UK, 1998: Cambridge University Press and Sky Publishing Corp.

Sinnott, R. W., and M. A. C. Perryman, *Millennium Star Atlas*, 3 vols., Cambridge, MA, 1997: Sky Publishing Corp. and European Space Agency.

Strong, R. A., and R. W. Sinnott, *Sky Atlas 2000.0 Companion*, 2nd edition, Cambridge, MA, 2000: Sky Publishing Corp. and Cambridge University Press.

Tirion, W., B. Rappaport, W. Remaklus, M. Cragin, and E. Bonanno, *Uranometria 2000.0 Deep Sky Atlas* and *Deep Sky Field Guide*, 3 vols., Richmond, VA, 2001: Willmann-Bell.

Tirion, W., and R. W. Sinnott, *Sky Atlas 2000.0*, 2nd edition, Cambridge, MA, 1998: Sky Publishing Corp. and Cambridge University Press.

Helpful Links

Astronomical League, *The Herschel 400 Club Observing List*, astroleague.org/al/obsclubs/herschel/hers400.html

Mason, B. D., G. L. Wycoff, and W. I. Hartkopf, *Washington Double Star Catalog*, ad.usno.navy.mil/wds/wds.html

Night Sky magazine, NightSkyMag.com

Sky & Telescope magazine, SkyandTelescope.com

Guide to Constellations

After each constellation's name is its standard three-letter abbreviation (in parentheses). This is followed by the chart(s) on which the constellation, or at least an important part of it, is shown. Note that Serpens has two sections.

Andromeda (And), 3, 72
Antlia (Ant), 37, 38, 39
Apus (Aps), 40, 50, 60, 70
Aquarius (Aqr), 75, 76
Aquila (Aql), 64, 65, 66
Ara (Ara), 58, 60, 69
Aries (Ari), 4
Auriga (Aur), 12
Bootes (Boo), 42, 44, 53, 55
Caelum (Cae), 18
Camelopardalis (Cam), 11, 21, 31
Cancer (Cnc), 24
Canes Venatici (CVn), 32, 43
Canis Major (CMa), 27
Canis Minor (CMi), 25
Capricornus (Cap), 66, 68, 77
Carina (Car), 28, 30, 39, 40
Cassiopeia (Cas), 1, 3, 72
Centaurus (Cen), 48, 59
Cepheus (Cep), 71, 73
Cetus (Cet), 6, 7
Chamaeleon (Cha), 30, 40, 50, 60
Circinus (Cir), 48, 50, 59, 60
Columba (Col), 18, 29
Coma Berenices (Com), 43, 45, C
Corona Australis (CrA), 69
Corona Borealis (CrB), 53, 55
Corvus (Crv), 47
Crater (Crt), 36
Crux (Cru), 38, 49, 50

Cygnus (Cyg), 62, 73
Delphinus (Del), 64
Dorado (Dor), 18, 20, D
Draco (Dra), 41, 42, 51, 52, 61
Equuleus (Equ), 75
Eridanus (Eri), 6, 16, 17, 19
Fornax (For), 6, 8, 17, 19
Gemini (Gem), 23, 25
Grus (Gru), 78, 79
Hercules (Her), 52, 54
Horologium (Hor), 8, 19
Hydra (Hya), 24, 26, 36, 37, 39, 46, 47
Hydrus (Hyi), 10, 20
Indus (Ind), 68, 79, 80
Lacerta (Lac), 72, 73
Leo (Leo), 34, 35
Leo Minor (LMi), 33, 35
Lepus (Lep), 16
Libra (Lib), 46, 57
Lupus (Lup), 48, 59
Lynx (Lyn), 22, 23
Lyra (Lyr), 63
Mensa (Men), 20, 30, D
Microscopium (Mic), 68, 79
Monoceros (Mon), 25, 27
Musca (Mus), 40, 50
Norma (Nor), 58, 59
Octans (Oct), 10, 60, 70, 80
Ophiuchus (Oph), 54, 56

Orion (Ori), 14, 16, B
Pavo (Pav), 70
Pegasus (Peg), 74, 75
Perseus (Per), 2, 13
Phoenix (Phe), 9, 78
Pictor (Pic), 18, 29
Pisces (Psc), 5, 74
Piscis Austrinus (PsA), 76, 77, 79
Puppis (Pup), 26, 27, 28, 29
Pyxis (Pyx), 26, 28, 39
Reticulum (Ret), 19, 20
Sagitta (Sge), 64
Sagittarius (Sgr), 66, 67, 68, 69
Scorpius (Sco), 56, 58
Sculptor (Scl), 7, 9, 78
Scutum (Sct), 67
Serpens Caput (Ser), 55
Serpens Cauda (Ser), 65, 67
Sextans (Sex), 36, 37
Taurus (Tau), 14, 15, A
Telescopium (Tel), 68, 69
Triangulum (Tri), 2, 4
Triangulum Australe (TrA), 60
Tucana (Tuc), 10, 80
Ursa Major (UMa), 31, 32, 33, 43
Ursa Minor (UMi), 41, 51
Vela (Vel), 28, 39
Virgo (Vir), 44, 45, 46, 47, C
Volans (Vol), 30
Vulpecula (Vul), 62, 64

Greek Letters on Charts

By tradition, the bright stars in each constellation are marked with lowercase Greek letters. A constellation's most brilliant star is often called α, but there are exceptions. For example, Gemini's brightest star is β and that of Octans is ν.

α	Alpha	ε	Epsilon	ι	Iota	ν	Nu	ρ	Rho	ϕ	Phi
β	Beta	ζ	Zeta	κ	Kappa	ξ	Xi	σ	Sigma	χ	Chi
γ	Gamma	η	Eta	λ	Lambda	o	Omicron	τ	Tau	ψ	Psi
δ	Delta	θ	Theta	μ	Mu	π	Pi	υ	Upsilon	ω	Omega

Charts 1–10

Right Ascension 0ʰ to 3ʰ

This section is highest in the night sky around these times:

Evening: **November & December**

Midnight: October

Morning: August & September

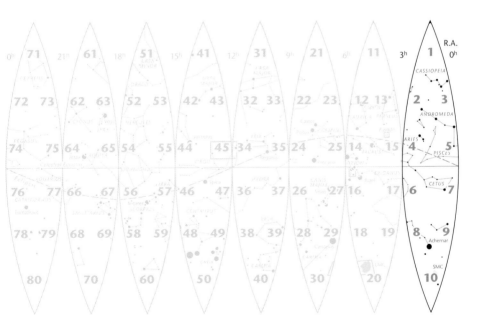

Chart Legend

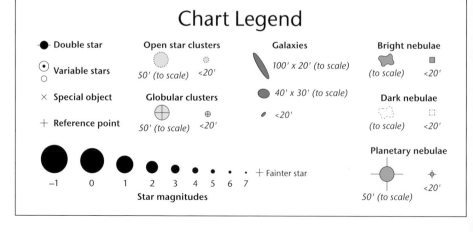

—●— Double star

⊙ Variable stars
○

✕ Special object

＋ Reference point

Open star clusters

50' (to scale)　　<20'

Globular clusters

⊕ 50' (to scale)　　⊕ <20'

Galaxies

100' x 20' (to scale)

40' x 30' (to scale)

● <20'

Bright nebulae

(to scale)　　■ <20'

Dark nebulae

(to scale)　　<20'

Planetary nebulae

50' (to scale)　　◇ <20'

−1　　0　　1　　2　　3　　4　　5　　6　　7　　＋ Fainter star

Star magnitudes

1

10ʰ 11ʰ 12ʰ 13ʰ 14ʰ 15ʰ 16ʰ 17ʰ

18ʰ
19ʰ
20ʰ
21ʰ

γ

2715
2655

R

ε

35

URSA
MINOR

δ

λ

North Celestial Pole

DRACO

2336
2300
2268 2276

α
Polaris

CAMELOPARDALIS

75

κ

2146

188 AR

S(c)

Σ460

CEPHEUS

U RX

El

ρ

SS

X

Errai

γ

31

π

1560

47

49

23

γ
IC 356

40

+70°

40

IC 342

50

Cr 463

RZ

V393
48

42

ο

vdB 8

ι

ω

31

7822

Ced 214

Σ3053

CASSIOPEIA

ψ

Tr 3

Σ163

6

IC 289

637

9

7790

IC 1871
Cr 34
IC 1795
896

ε

559

κ

12

+60°

Cr 33
1027
IC 1805
IC 1848
Mrk 6

654

663
V770
659 M103

381

225

136

IC 59

129

(c)
β Caph

γ

IC 63

Stock 2

δ
Ruchbah

υ²

IC 10 TV
vdB 1

Σ3062

η

884 869

χ

436
φ

υ¹

957

Sh2-188

457
V465

η

Schedar

Tr 2
DM
9
KK

281

α

T

PERSEUS

Double Cluster

2ʰ 1ʰ

71

4

1ʰ   0ʰ   23ʰ

1   **71**   **72**

637

559

κ   9   6   M52   4   7538 7510
381   12   Bubble   IC 1470   W°   KY
225   *136*   *7790*   Nebula,   7635   7380
129   (c)   7635   Sh2-157   V509
IC 59   β Caph   *V*   *1*
V770   IC 63   γ   IC 10 ° TV   Σ3062   τ   Bradley   LACERTA
M103   Ruchbah   υ²   vdB 1   ρ   3077
χ   δ   *436*   υ¹   *7789*   CASSIOPEIA
φ V465   η   σ
Sh2-188   *457*
281   Schedar   T   SV   +50°
θ   μ   α   λ   3
ζ   R   18   7   8   11
TU°   (c)   7686   Z   4
ν   ξ   λ
147   ψ   Blue
185   o   κ   ι   Snowball
51   278   π   7662   7640   +40°
49   φ   RV   22
ξ   26   14
χ   ω   41   Andromeda
υ   Galaxy   M110   206
M31   M32   θ   Σ3050
ν   EG   32   R°   ρ
μ   σ   V343°   +30°
ANDROMEDA   78
404   π
Mirach   β   δ   α
M33   82   σ   Alpheratz   PEGASUS
PISCES   315   28   85   7741
91   τ   ε   Σ24   ψ
68   65
υ   RT°
(c)   φ   ζ

1ʰ   **5**   0ʰ

4

2

3ʰ   2ʰ

PERSEUS

TRIANGULUM

β

δ
γ

15
W
R
Σ285
925
7
ε

21

1333

M33

vdB 16

+30°

6

12

α

39

Σ239

784

10

41
35
33

1156

672

IC 1727

64

30

14

10

R

Hamal
α

λ

τ
ζ

ν

κ

ε

η

ARIES

Sheratan

63

β

107

+20°

δ

μ

θ

γ

772

Mesartim

50°

ρ

ι

π
T

M74

U

η

σ
ο

40°

660

38

π

+10°

ξ

821

25 Ari

λ

μ

ο

ξ²

ξ¹

Σ138

864

676

ν

ν

Menkar
α

II Zw 5

γ

α

ξ

Alrescha

1073

Σ186

1055
δ

0°

94

Σ330

M77

69

60

1087

R

75
936
70

CETUS

Mira
ο

ERIDANUS

3ʰ   2ʰ

15

6

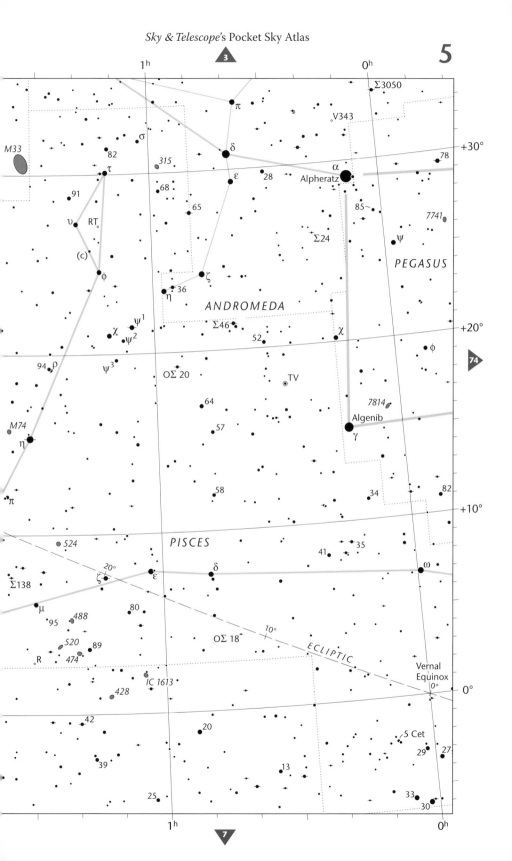

1ʰ   0ʰ

Σ3050

π

○V343

M33

σ

82

315

δ   78

+30°

τ   α

91   68   ε   28   Alpheratz

85

65   7741

υ   RT₀   Σ24   ψ

(c)   ψ

φ   ζ   *PEGASUS*

36

η   *ANDROMEDA*

χ   ψ¹   Σ46   χ   +20°

ψ²   52   φ

94 ρ   ψ³   ▶ 74

OΣ 20   TV

64   7814

M74   57   Algenib

η   γ

58   34   82

π   +10°

PISCES

524   35

41   ω

20°   δ

Σ138   ζ   ε

μ

95   488   80

520   89   OΣ 18   *ECLIPTIC*

R   474   10°

IC 1613   Vernal

Equinox

428   0°   0°

42   20   5 Cet

39   29   27

13

25   33

30

6

4

3ʰ

2ʰ

α
Menkar

II Zw 5

α Alrescha

ξ

R

γ

Σ186

1073

1055

δ

69

60

1087

M77

°R

70

Σ330

75

936

0°

Mira

ο

CETUS

779

1022

584

596

615

ρ³

ρ¹

1084

1052

Baten
Kaitos

ρ²

RR

1042

988

Azha

ζ

χ

η

17

ε

ρ

Z°

°U

720

π

AS

50

σ

τ

ERIDANUS

°UV

τ¹

57

1300

υ

48

−20°

1232

τ²

h3511

908

578

τ⁴

56

15

°AA

1187

ε

τ³

4

γ¹

κ

ζ

1302

1255

R(c)°

1201

FORNAX

γ²

ω

ε

ν

τ

613

α

1097

ι¹

π

μ

ι²

−30°

π

R(c)°

β

φ

λ² λ¹

Fornax Dwarf

η¹

3ʰ

2ʰ

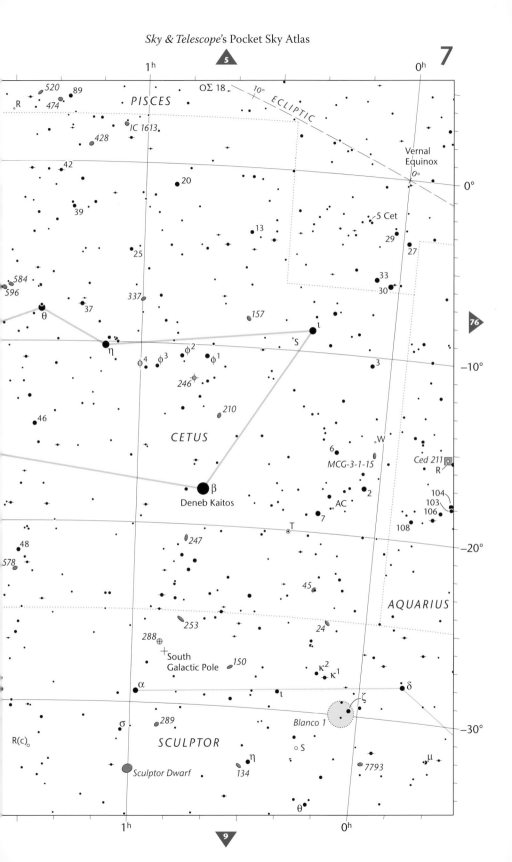

1ʰ

0ʰ

OΣ 18

ECLIPTIC

10°

PISCES

520

89

R

474

IC 1613

428

42

20

39

Vernal Equinox

0°

0°

5 Cet

29

27

13

33

25

30

337

584

596

θ

37

157

η

φ⁴ φ³ φ² φ¹

246

ι

S

3

−10°

46

210

CETUS

6

MCG-3-1-15

W

Ced 211

R

β

2

104

Deneb Kaitos

AC

103

106

T

7

108

247

−20°

48

578

45

AQUARIUS

253

24

288

150

κ²

κ¹

South
Galactic Pole

α

ι

δ

ζ

σ

289

Blanco 1

−30°

R(c)

SCULPTOR

S

μ

η

Sculptor Dwarf

134

7793

θ

76

3ʰ 2ʰ

ERIDANUS

κ

ε

τ⁴ 1187
15 τ³ 4 γ¹
ζ R(c)°
1255 1201 ν
1302 ω π 613
ε γ² μ τ
ι² ι¹ π
1097 π

α FORNAX

β φ
λ² λ¹
1344 Fornax Dwarf
η³ η¹
η²

1350 χ² χ¹ ψ 986
1374 1379
1380 χ¹ 625
1387 1326
1427 1399 1317 ι φ
1404 χ³ 1316 = Fornax A χ
1386 1365 s ψ

h Acamar
θ δ
y 1291
EU κ
−40° h3556 ERIDANUS φ χ
1411 685
1448 q²
R q¹
1493 °U 1433 ι
1494 T° η Δ5 = p Eri
1527 Achernar
HOROLOGIUM ζ α
IC 1954 1249
γ 1261
1515 h3592
TW
1566 1549 V°
1617 1553 1533 μ λ
1596 1546 α
α 1574 1543 RETICULUM
DORADO ε ν
ι δ γ
κ ζ² ζ¹ β

−30° −50° −40° −50°

19 20 10

4ʰ 3ʰ 2ʰ

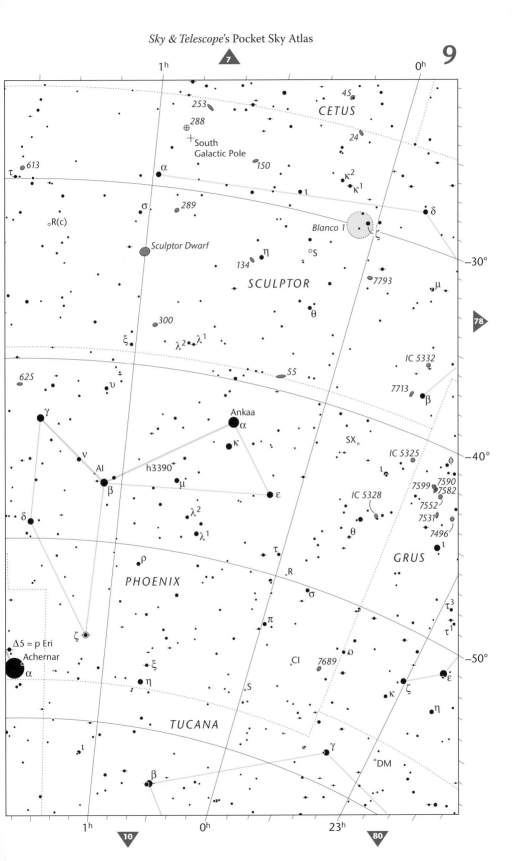

1ʰ

CETUS

45

253

288

⊕

✛ South
Galactic Pole

150

24

κ²
κ¹

ι

0ʰ

τ

613

α

σ

289

Blanco 1

ζ

δ

−30°

R(c)

Sculptor Dwarf

134

η

°S

7793

μ

SCULPTOR

θ

▶ 78

300

ξ

λ² • λ¹

55

IC 5332

7713

β

625

υ

γ

Ankaa
α

κ

SX °

IC 5325

φ

−40°

ν

Al

β

h3390

μ

ι

ε

IC 5328

7599   7590
7582
7552
7531
7496

δ

λ²

λ¹

θ

ι

ρ

τ

R

GRUS

PHOENIX

σ

τ³

π

τ¹

Δ5 = p Eri
Achernar
α

ζ

ξ

η

CI

7689

υ

S

−50°

ζ

ε

κ

η

TUCANA

γ

°DM

ι

β

23ʰ

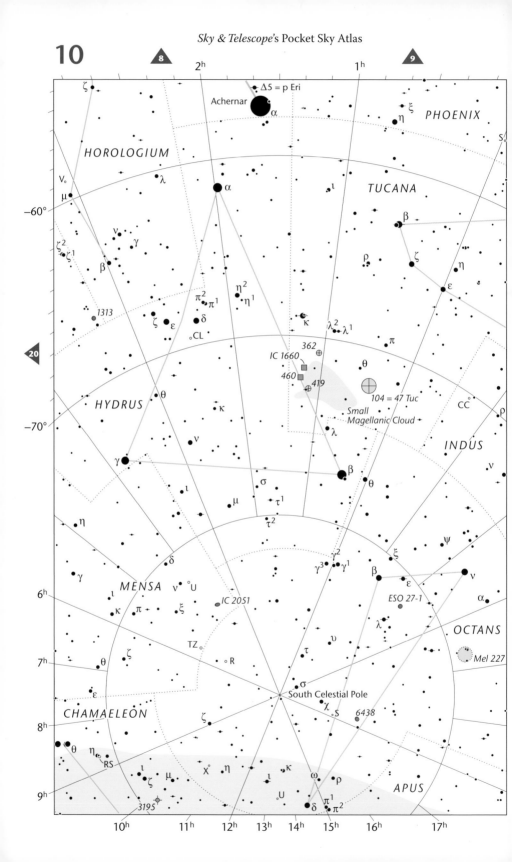

Charts 11–20

Right Ascension 3ʰ to 6ʰ

This section is highest in the night sky
around these times:

Evening: **January**

Midnight: November & December

Morning: October

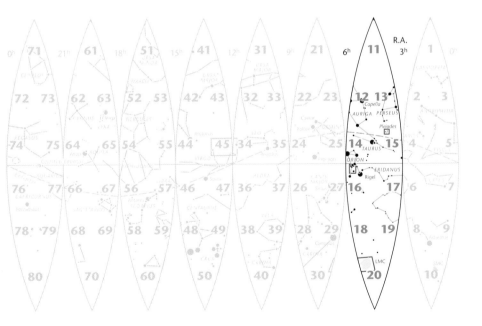

Chart Legend

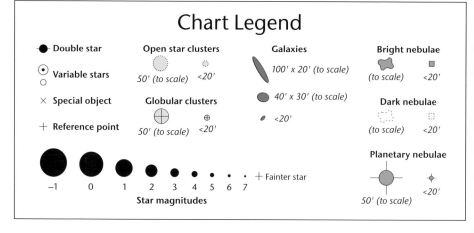

Double star

Variable stars

× **Special object**

+ **Reference point**

Open star clusters

50' (to scale)   <20'

Globular clusters

50' (to scale)   <20'

Galaxies

100' x 20' (to scale)

40' x 30' (to scale)

<20'

Bright nebulae

(to scale)   <20'

Dark nebulae

(to scale)   <20'

Planetary nebulae

50' (to scale)   <20'

−1   0   1   2   3   4   5   6   7   + Fainter star

Star magnitudes

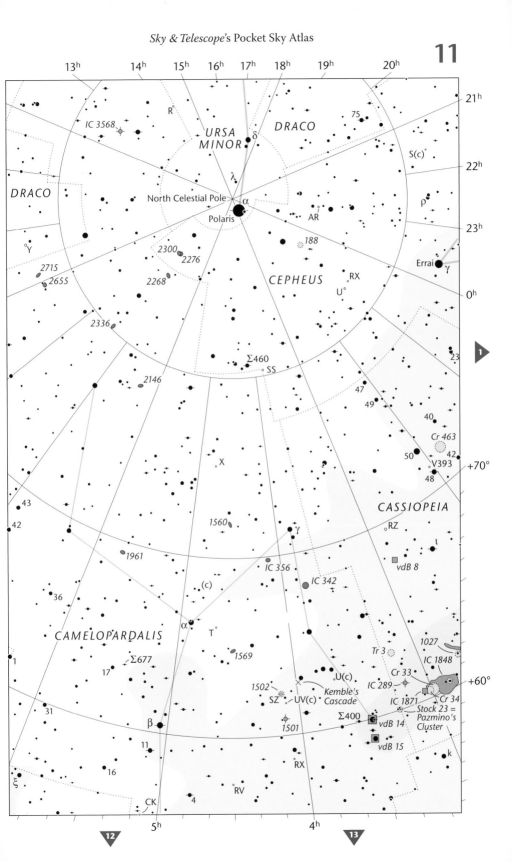

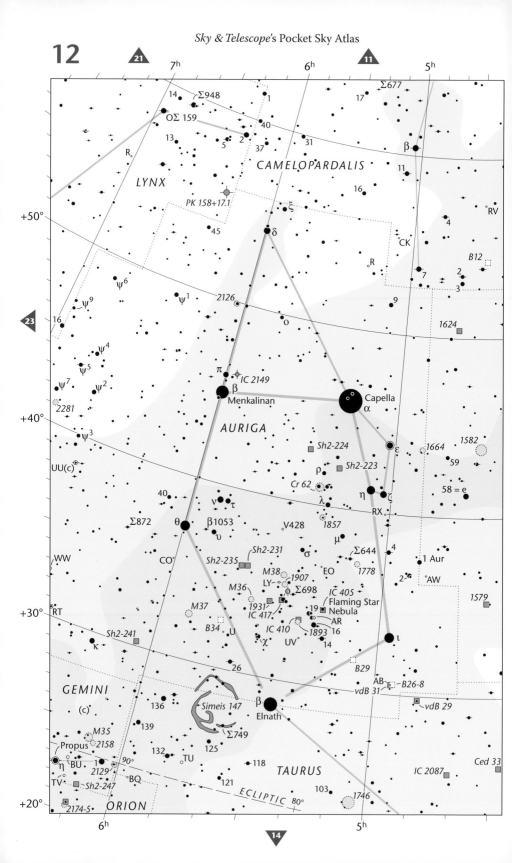

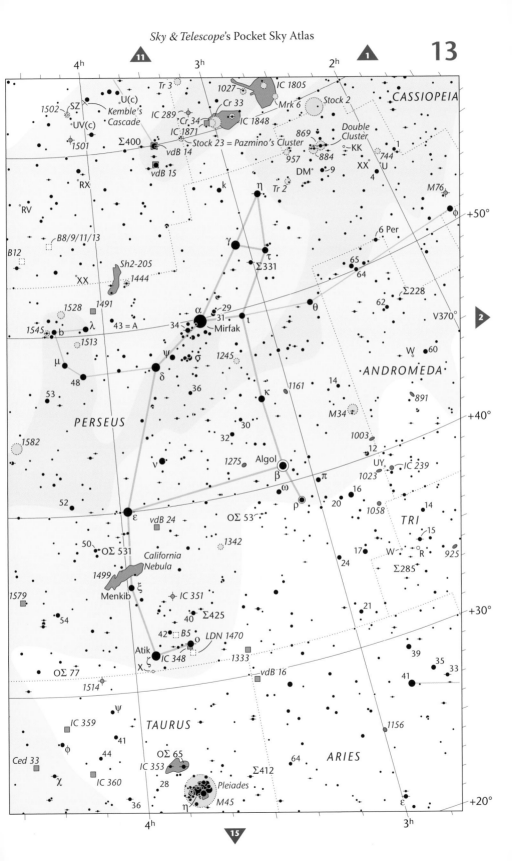

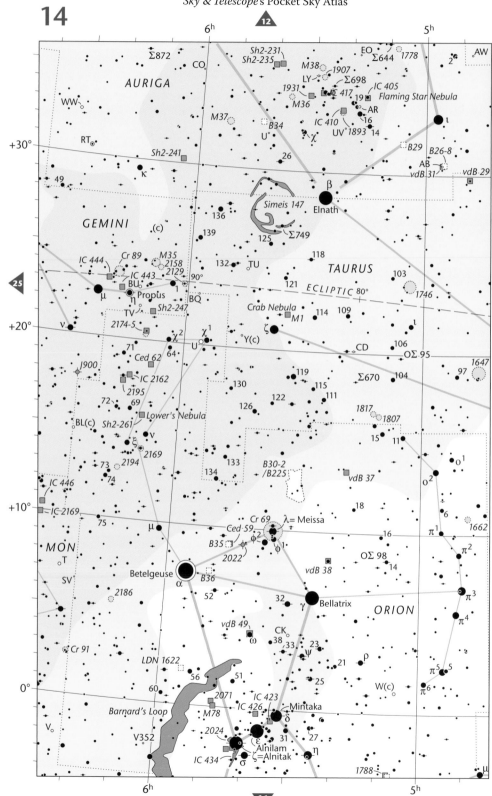

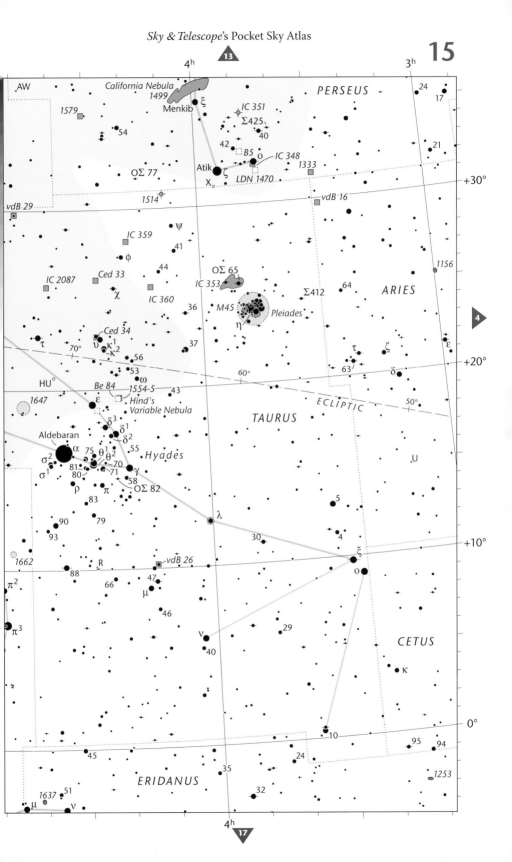

AW

California Nebula
1499
Menkib   ξ

1579

54

PERSEUS

24

17

IC 351

Σ425

40

42

21

OΣ 77

Atik
χ₀

B5
ο   IC 348
ζ   LDN 1470

1333

1514

+30°

vdB 29

ψ

IC 359

41

φ

vdB 16

vdB 16

IC 2087

Ced 33

χ

44

IC 360

OΣ 65

IC 353

36

M45

η

Σ412

64

1156

ARIES

Pleiades

Ced 34

τ   70°   υ   κ¹   κ²

56

37

τ   ζ

ε

+20°

53   ω

HU°

Be 84   1554-5

43

63

δ

60°

50°

1647

ε

Hind's
Variable Nebula

ECLIPTIC

TAURUS

δ³

δ¹

Aldebaran
α   75
σ²   81   θ¹   θ²
σ¹   80   71

δ²

55   *Hyades*

U

ρ   π   OΣ 82

83

90

79

58   γ

70

5

4

93

λ

30

+10°

1662

88   R

vdB 26

ξ

ο

π²

66

47

μ

46

ν

40

29

CETUS

κ

π³

10

0°

95   94

45

24

35

1253

ERIDANUS

32

1637   51
μ   ν

6^h 5^h

2186

vdB 49   CK
ω 38
33   23
ψ
ORION   25   21   ρ
π 5
5
LDN 1622   π 6
W(c)
56   51
2071   IC 426
60   IC 423   Mintaka
M78   δ
0°   *2024*   22
Barnard's Loop   31   27
ε   η
1788   68   *1637*   51
V352   Alnilam   μ
ζ = Alnitak   66   *1700*
IC 434   σ   β   ω
45   *1981*   Cursa
42   S
M43   *Witch Head Nebula*
2183   *2182*   *Orion Nebula*   ψ
2185   *2170*   M42   IC 2118
γ   1999   ι
IC 430   τ   Σ649
2215   49   29   λ
7   υ   Rigel   β
MONOCEROS   55   Saiph   63
2149   2   κ   *ERIDANUS*
3
−10°
ν   RX   64
IC 418   λ   ι   53
8   κ
MCG-2-15-11   R(c)   Hind's
θ   η   ζ   *1832*   Crimson Star
17   μ   60
58
Arneb   54
β 321   α
S476
β   Mirzam   19   Nihal
2204   β   h3750
δ   *1964*   T
−20°   *2196*   γ   ε
1744
ξ^1 S   *2139*   M79
LEPUS   (c)

2217   ν^1
CANIS   ν^2
MAJOR
ζ   *COLUMBA*   ζ
−30°   Furud   σ   *1679*
μ   *CAELUM*
λ   δ   λ   Phact   T
2188   *2090*   α   o

6^h 18   5^h

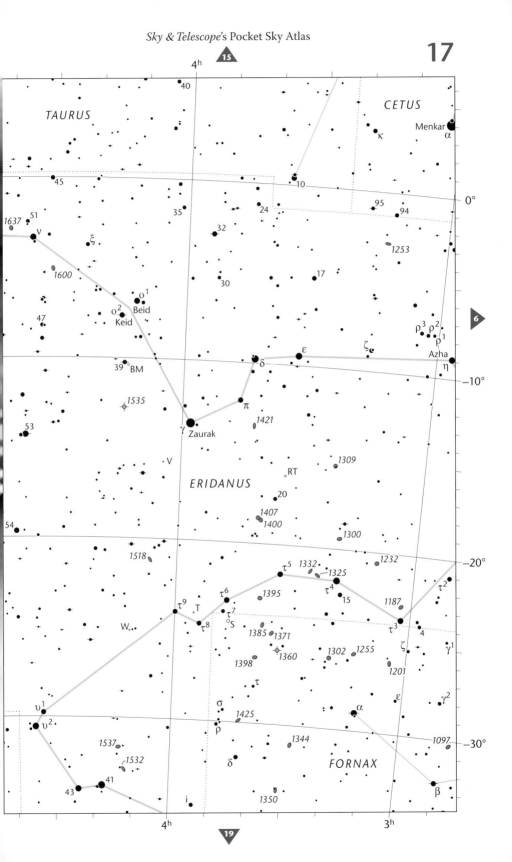

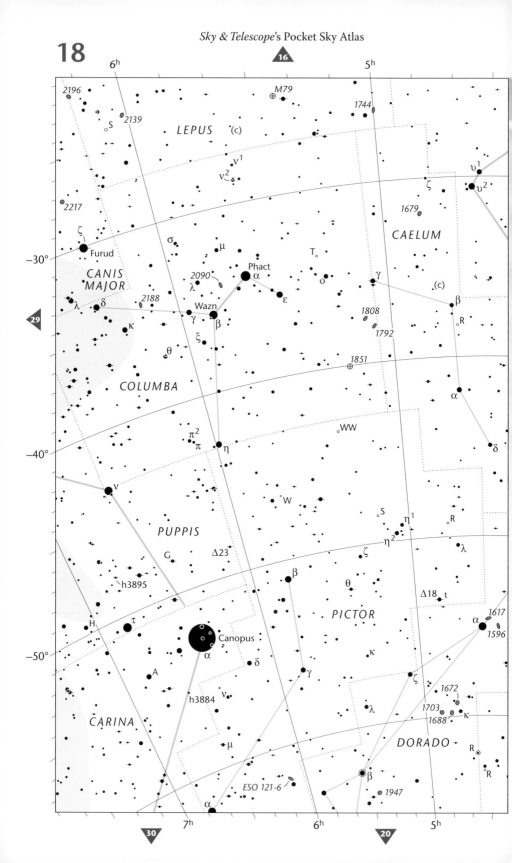

18

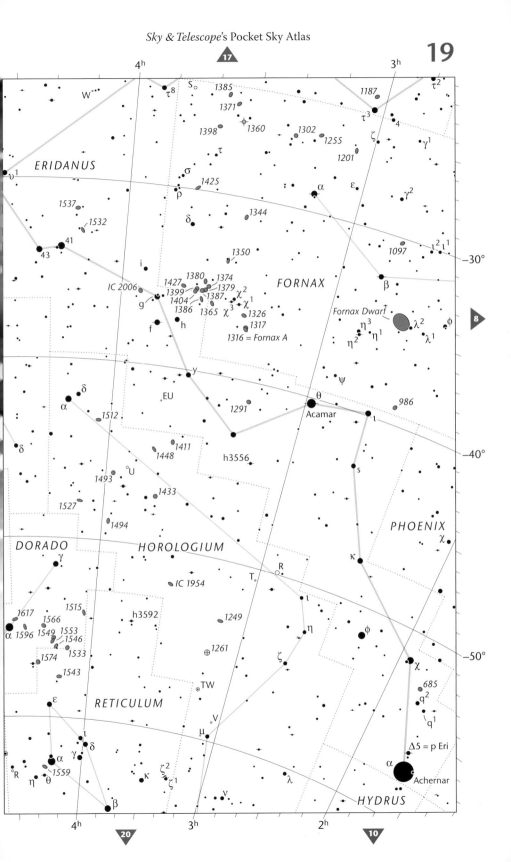

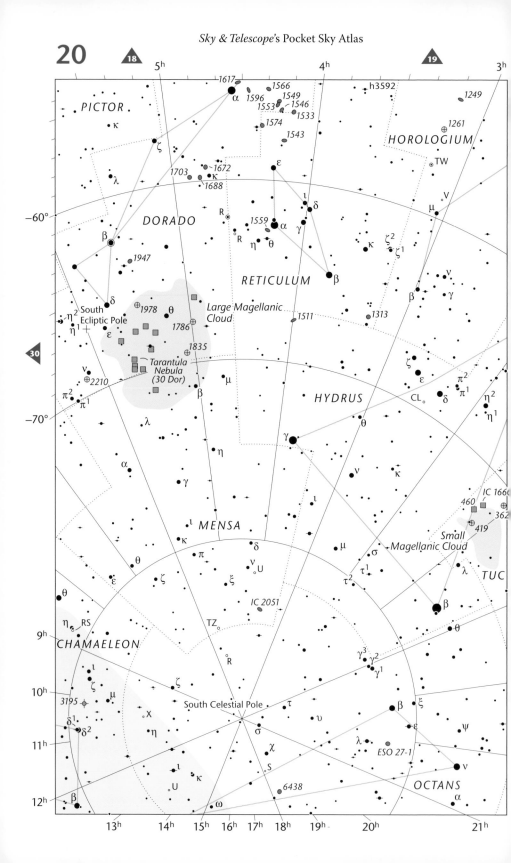

20

Charts 21–30

Right Ascension 6^h to 9^h

This section is highest in the night sky around these times:

Evening: **February & March**

Midnight: January

Morning: November & December

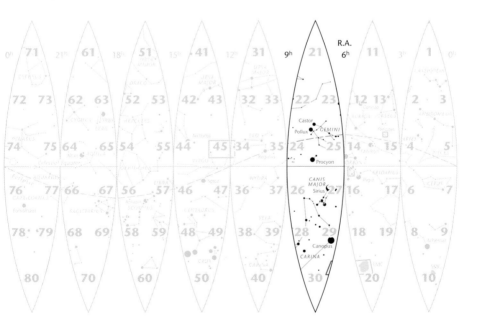

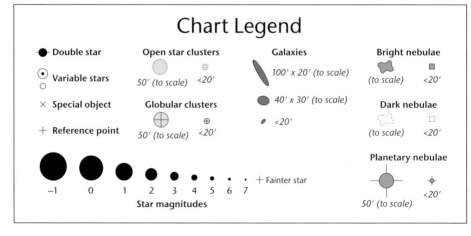

Chart Legend

Double star

Variable stars

× **Special object**

+ **Reference point**

Open star clusters

50' (to scale)   <20'

Globular clusters

50' (to scale)   <20'

Galaxies

100' x 20' (to scale)

40' x 30' (to scale)

<20'

Bright nebulae

(to scale)   <20'

Dark nebulae

(to scale)   <20'

Planetary nebulae

50' (to scale)   <20'

−1   0   1   2   3   4   5   6   7   + Fainter star

Star magnitudes

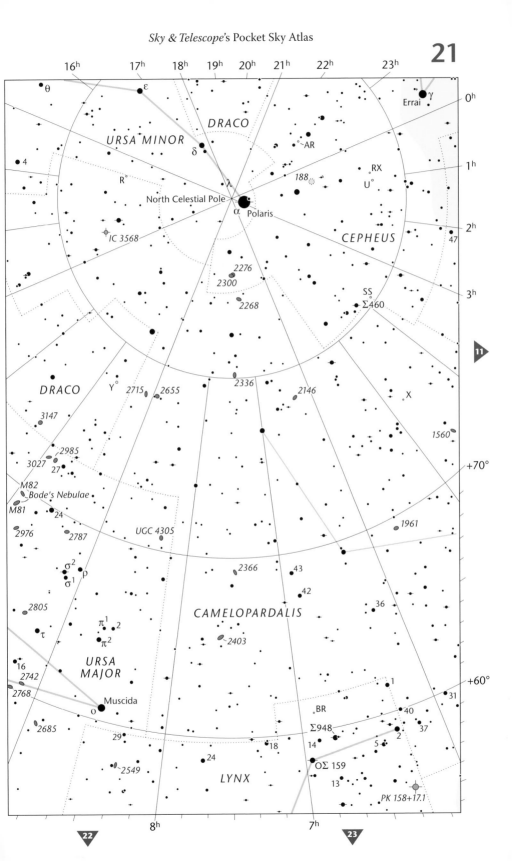

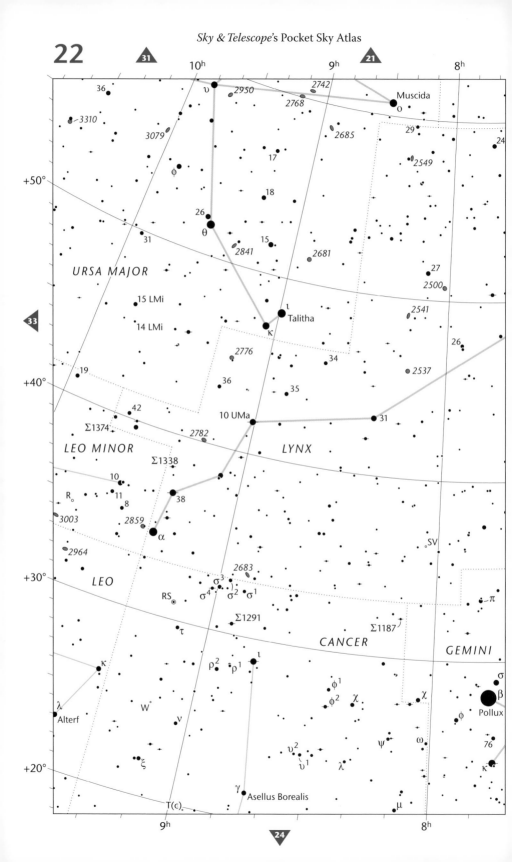

31

21

33

24

10ʰ

9ʰ

8ʰ

36

·3310

2950

2742

2768

Muscida

o

29

24

3079

17

2685

2549

+50°

18

26

2681

27

θ

2500

31

15

2841

ι Talitha

URSA MAJOR

κ

2541

15 LMi

26

14 LMi

2776

34

2537

19

36

35

+40°

42

10 UMa

31

Σ1374

2782

LEO MINOR

LYNX

Σ1338

10

SV

11

38

R

8

2859

3003

α

2964

SV

π

+30°

LEO

σ³

2683

RS

σ⁴

σ²

σ¹

τ

Σ1291

Σ1187

CANCER

GEMINI

κ

ρ²

ι

σ

λ

ρ¹

β

Alterf

φ¹

χ

Pollux

W

φ²

χ

ν

φ

ψ

ω

76

υ²

+20°

ξ

υ¹

λ

κ

μ

T(c)

γ Asellus Borealis

9ʰ

8ʰ

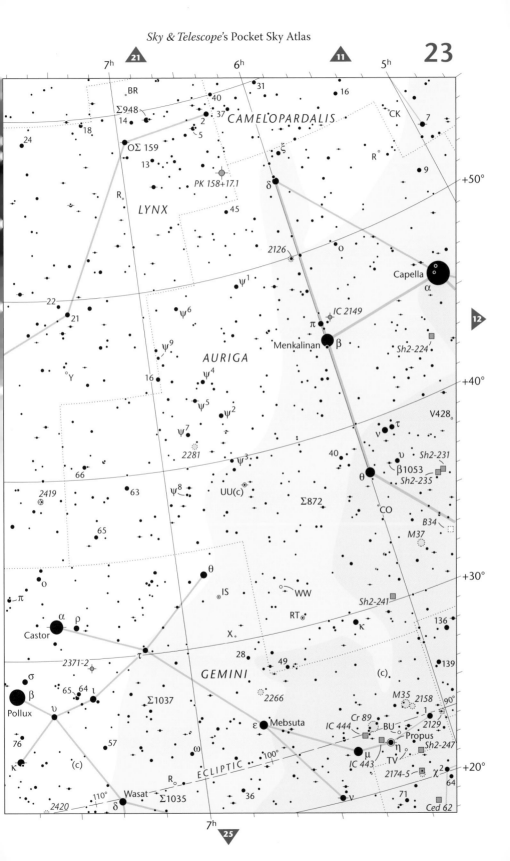

9ʰ   8ʰ

LYNX

2683

σ²

σ⁴   σ³   σ¹

RS⊙   *LEO*   Σ1187

Σ1291   *GEMINI*

+30°

τ

ι   σ

ρ²   ρ¹   φ¹   β

κ   χ   Pollux

φ²   χ   φ

W∘   ψ   ω

ν   76

υ²   λ   κ

Alterf   υ¹

λ   *CANCER*

2903   ξ   μ

+20°   η

γ Asellus Borealis

T(c)   Beehive   120°

ε   Cluster M44   85

130°   δ   ζ   81

ECLIPTIC   θ

140°   Asellus

π   81   o   Australis   °V

27   8

α   R

Acubens 60   ⊙ M67   11

κ   ∘RT

ω   VZ

∘PK 219+31.1   β

2775

+10°

ζ   ε

ω   ρ   δ

θ   S   σ   14   ζ

η

0°

28

ζ

HYDRA   14

C   27

23

9ʰ   8ʰ

+30°

+20°

35

+10°

0°

7^h

o
-π
Castor
ρ
α
θ
IS
WW
AURIGA
Sh2-241
+30°

M37

2371-2
σ
β
64
ι
65
Pollux
υ
76
57
κ
(c)
Σ1037
2266
28
49
X
RT
κ
136
139
(c)
ε Mebsuta
ω
Cr 89
M35 2158 2129
IC 444 IC 443 BU 90°
μ η Propus BQ
TV Sh2-247
100°
ECLIPTIC
R
Wasat
110°
Σ1035
δ
36
ν
71
2174-5 χ 2
64
U
+20°
2420
63°
2392=Eskimo Nebula
ζ Mekbuda
56
GEMINI
Ced 62
IC 2162
2195
81
74
BN
68
λ
51
45
2304
26
γ
Alhena
W
Σ932
BL(c)
72 69
Lower's Nebula
Sh2-261
73
74
ν
ξ 2169
2194
2355
2395
PK 205+14.1
Medusa Nebula
6
1
30
38
ξ
ORION
75
μ
+10°
11
R(c)
γ
ε
OΣ 170
V
S
Gomeisa
β
η
Σ1126
Procyon
α
δ³
δ²
δ¹
CANIS MINOR
B37-9
2264
Cone Nebula
2247
15
2245
IC 446
IC 2169
Hubble's Variable Nebula
2261
17
2254
13
2251
IC 448 T
SV
2186
RV
Cr 111
Cr 97
2237-8/46
Cr 106
2252
2244
Cr 107
Rosette Nebula
Cr 96
Cr 91
18
V505
Sh2-282
2324
2301
2282
Do-25
0°
21
δ
MONOCEROS
V
2286
25
IC 466
(c)
20
19
2311
10
2232
2183 2182
2185
γ

7^h

◀ 14

9h

°S

θ

0°

14

HYDRA

C

RT°

M48

8h

14

ζ

*CANIS
MINOR*

28

ζ

27

25

MONOCEROS

α

23

F

20

MCG-1-24-1

24

27

FK

°T

RV

2506

−10°

37

26

6

12

9

2811

AK

19 2539

20

2525

5

Mel 71

9

2423

2438

2 M46

2479

6 2440

2509

16

V407°

2421

−20°

G

2835

2613

PYXIS

IL

2784

θ

κ

δ γ

TY

λ

(c)

S°

ζ

2627

η

2566

2559

2571

2567

11

12

ρ

ζ

o

2482

M93

2467

2483

Ru 44

2527

2489

2533

2439

−30°

ε

ζ²

ζ¹

IC 2469

ε

α

2663

β

2579

Cr 185

q

h4063

AT

2546

AS

w

Ru 55

RS

PUPPIS

f

2451

c

d²

9h

8h

28

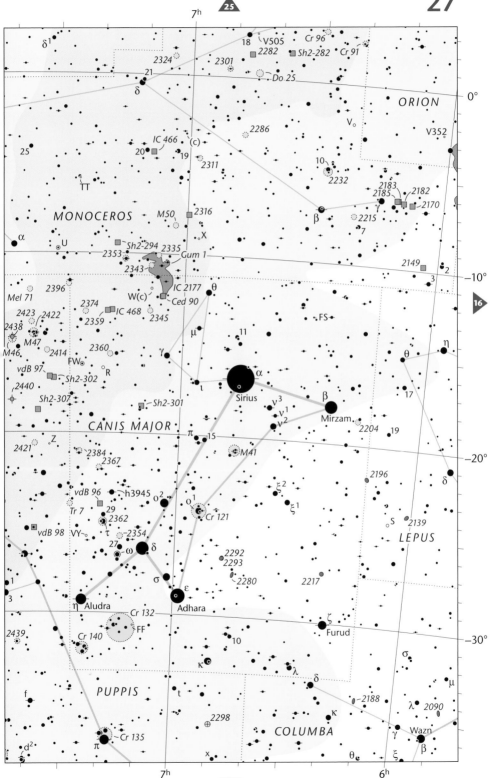

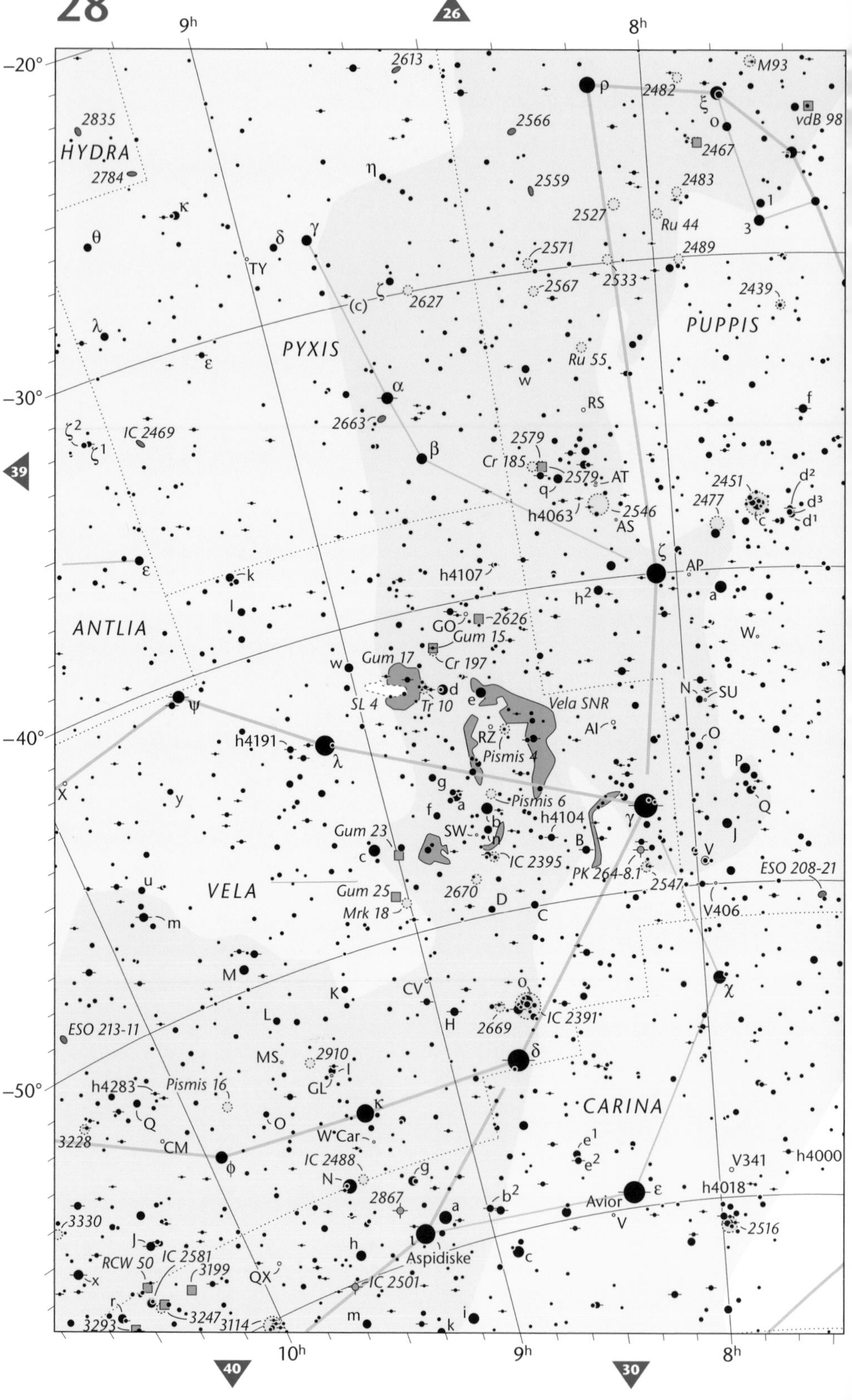
26
9h
8h
–20°
–30°
–40°
–50°
39
40
30
10h
9h
8h
HYDRA
PYXIS
PUPPIS
ANTLIA
VELA
CARINA
M93
2482
vdB 98
2613
2835
2566
2467
2784
2559
2483
2527
Ru 44
2571
2489
TY
2533
2567
2627
2439
Ru 55
RS
IC 2469
2663
2579
Cr 185
2579
AT
2451
2477
h4063
2546
AS
h4107
AP
GO
2626
Gum 15
Gum 17
Cr 197
SL 4
Tr 10
Vela SNR
SU
RZ
Pismis 4
AI
h4191
Pismis 6
h4104
Gum 23
SW
IC 2395
PK 264-8.1
2547
ESO 208-21
Gum 25
2670
Mrk 18
V406
CV
IC 2391
2669
ESO 213-11
MS
2910
GL
h4283
Pismis 16
3228
CM
W Car
IC 2488
V341
h4000
h4018
2867
Avior
2516
3330
Aspidiske
RCW 50
IC 2581
3199
QX
IC 2501
3293
3247
3114

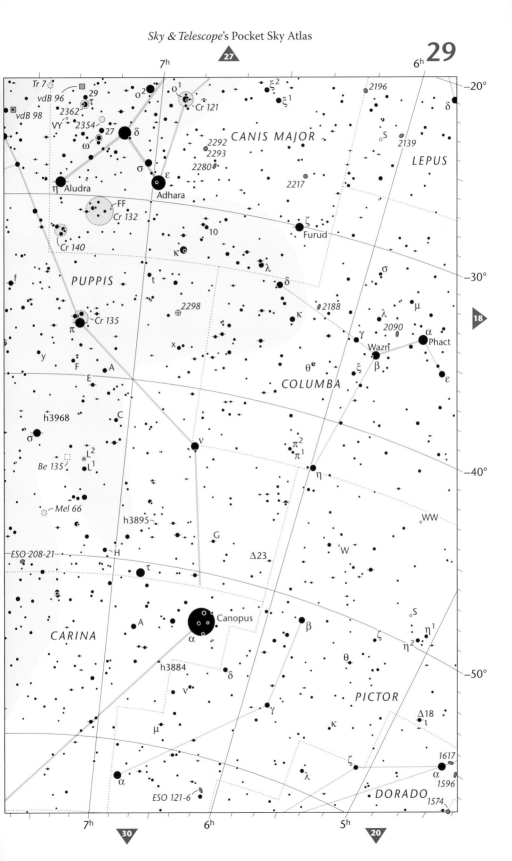

7ʰ

−20°

Tr 7
vdB 96
29
τ
o²
o¹
Cr 121
ξ²
ξ¹
2196
δ

vdB 98
2362
VY
2354
ω
27
δ

CANIS MAJOR

S
2139

LEPUS

2292
2293
2280

σ

η Aludra
ε
Adhara

2217

−30°

FF
Cr 132

ζ
Furud

Cr 140

10

κ

λ

δ

σ

PUPPIS

t

2298

κ

2188

μ

λ
2090
α
Phact

γ

Wazn
β

18

π
Cr 135

x

θᶜ
ξ

ε

y

F
A

COLUMBA

E

h3968

C

σ

Be 135
L²
L¹

ν

π²
π¹

η

−40°

Mel 66

WW

h3895

G

W

ESO 208-21

H

Δ23

τ

S

CARINA

A
α
Canopus
β

ζ
η¹
η²

θ

−50°

h3884
δ

ν

PICTOR

μ

γ

κ

Δ18
τ

α

ζ
λ

α
1617
1596

ESO 121-6

DORADO
1574

7ʰ ▽ **30**   6ʰ   5ʰ ▽ **20**

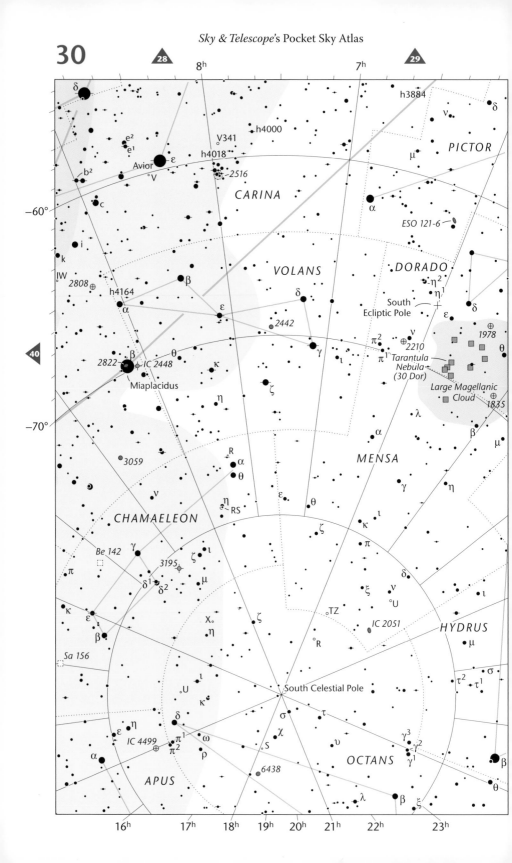

28
29
40

8ʰ
7ʰ

δ

h3884

ν
δ

PICTOR

μ

e²
e¹

Avior
°V
ε

V341

h4018

h4000

2516

α

CARINA

ESO 121-6

b²

c

−60°

i

k

VOLANS

DORADO

ο²
η²
η¹

δ

IW

2808

h4164

α
β

δ

ε

γ

South
Ecliptic
Pole

ε

ν

1978

θ

2442

π²

2210

θ

β
2822
IC 2448

Miaplacidus

κ

ζ

ι

π¹
ν

Tarantula
Nebula
(30 Dor)

Large Magellanic
Cloud

1835

η

λ

α

μ

−70°

3059

R
α
θ

ν

MENSA

β

γ

η

ε

θ

CHAMAELEON

η
RS

ζ

κ
ι

π

Be 142

γ

ι

3195

ζ

δ¹
δ²

μ

ε
ζ

θ

ζ

δ

ξ
ν
°U

ι

π

κ
β

X

η

TZ

IC 2051

HYDRUS

μ

Sa 156

R

°R

τ²
τ¹

σ

U
ι
κ

South Celestial Pole

σ

τ

δ
η

IC 4499

π¹
π²

ω

ρ

S

χ

υ

γ³
γ²
γ¹

β

α

6438

OCTANS

θ

APUS

λ

β
ξ

Charts 31–40

Right Ascension 9ʰ to 12ʰ

This section is highest in the night sky around these times:

Evening: April

Midnight: February & March

Morning: January

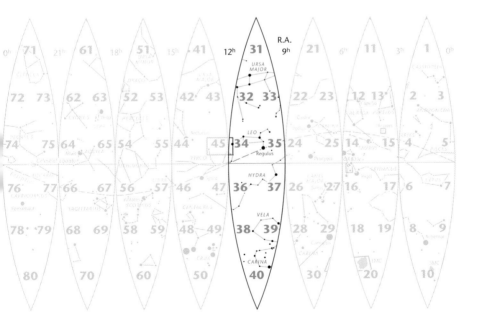

Chart Legend

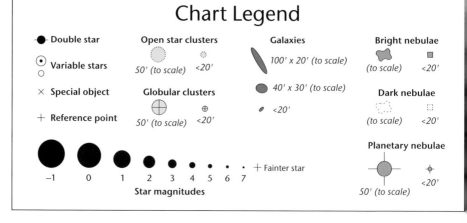

Double star

Variable stars

× **Special object**

+ **Reference point**

Open star clusters

50' (to scale)   <20'

Globular clusters

50' (to scale)   <20'

Galaxies

100' x 20' (to scale)

40' x 30' (to scale)

<20'

Bright nebulae

(to scale)   <20'

Dark nebulae

(to scale)   <20'

Planetary nebulae

<20'

50' (to scale)

−1   0   1   2   3   4   5   6   7   + Fainter star

Star magnitudes

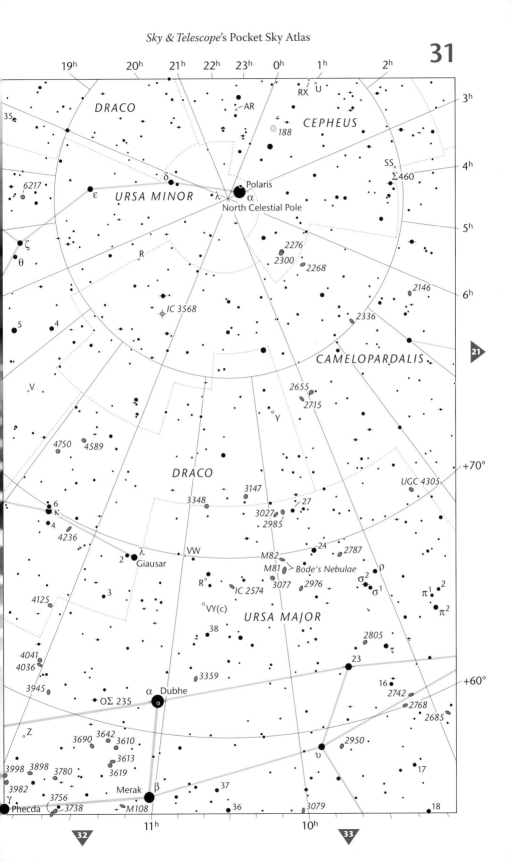

14ʰ   41   13ʰ   31   12ʰ   11ʰ

8

DRACO

4125

5322

5585

5204

4605

S

4041
4036
RY

Dubhe

5473
M101
5474

83

T

74

M40

Z

OΣ 235

α

3945

3642
3690
3610
3613

80 Alcor
ζ Mizar
78
ε
Alioth γ

δ
Megrez

3898 3780
3898
3982

3619

β

82

+50°

η
Alkaid

3998

M108
M97 Owl Nebula

URSA MAJOR

3738
3756
3729
3718

3631

43

24

21

5

M109
4102
3953

γ
Phecda

5195
M51
Whirlpool Galaxy

V

TU

4800

4157
4088
4085

4026

4100

3893

3583

Y(c)

3

M106
4346

4220
4217

4096
4144

3949
3877

χ

3726

M63
Sunflower Galaxy

4242

4449

ST

+40°

20

M94
4618
β
Chara
4485
4490

4051
4138
4013
4111
4143

3938

67

AZ

3675
56

ψ

α
Cor Caroli

2

5005
5033
14

6

4145
4151

CANES VENATICI

4244

57
3665

55

4214

Groombridge 1830
3941

3813

TV

61

Lalande 21185

4631
4656

4395

4203

+30°

37

T

4414

4062

Center of
Coma Galaxy Cluster

4889

31

4314

4150

4559
4448

4274
4278
4245
4136

Alula Borealis ν

North
Galactic Pole

γ

4251

Σ1633

14
4565
16
13 12

4725

4494

Σ1639

COM

21

3900

3912

Alula Australis ξ

3504

LEO

12ʰ   34

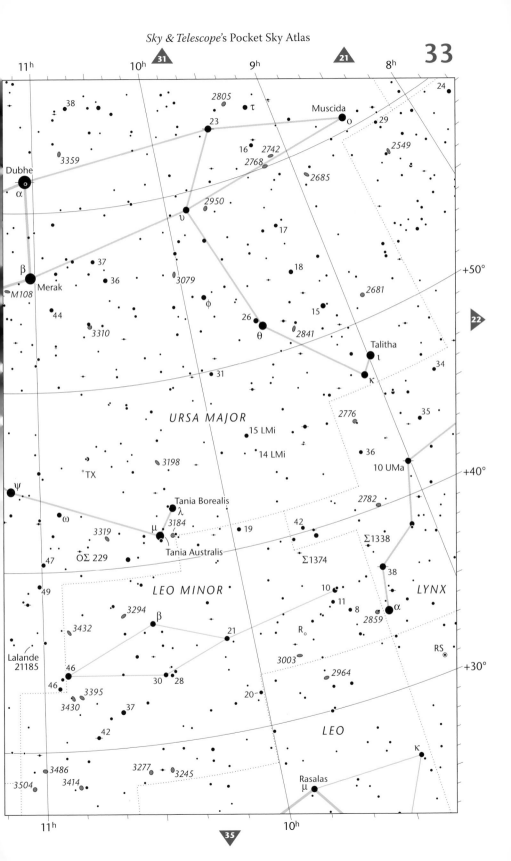

11ʰ　　　10ʰ　　　9ʰ　　　8ʰ

24

38

2805

τ

Muscida

ο

29

23

2549

16　2742

2768

3359

2685

Dubhe

α

2950

υ

17

+50°

18

β

37

2681

36

3079

Merak

M108

φ

15

22

44

26　　2841

3310

θ

Talitha

ι

31

34

κ

URSA MAJOR

2776

35

15 LMi

36

14 LMi

10 UMa

+40°

3198

TX

ψ

2782

Tania Borealis

λ

ω

μ　3184

42

Σ1338

3319

19

47

OΣ 229

Tania Australis

Σ1374

38

49

LEO MINOR

10

LYNX

3294

11

β

8

α

3432

21

2859

Lalande
21185

R

3003

RS

+30°

46

30　28

2964

46

20

3395

37

3430

LEO

42

κ

3486

3277　3245

3504

3414

Rasalas

μ

11ʰ　　　　　　　　10ʰ

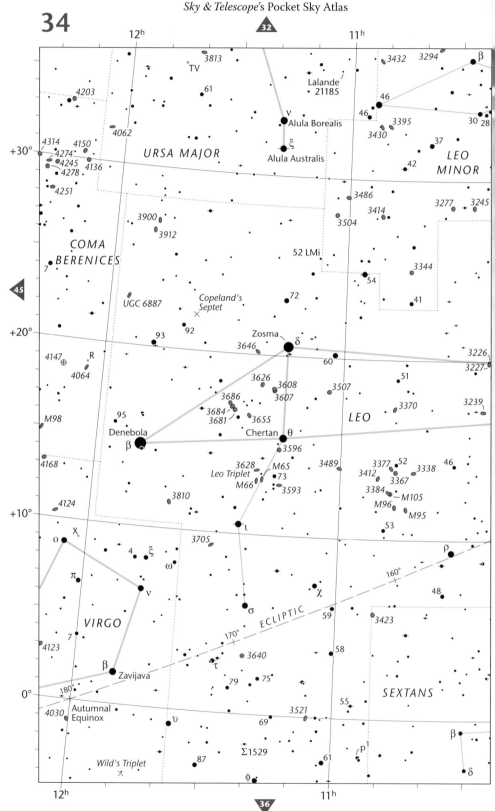

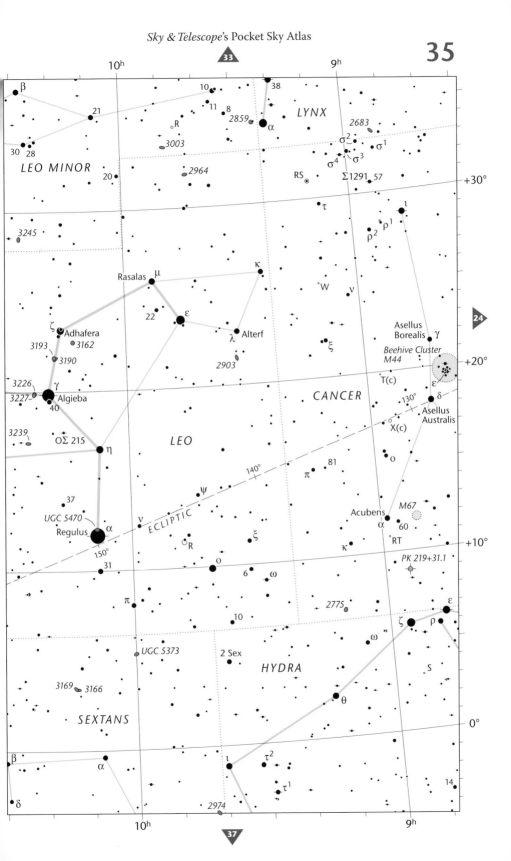

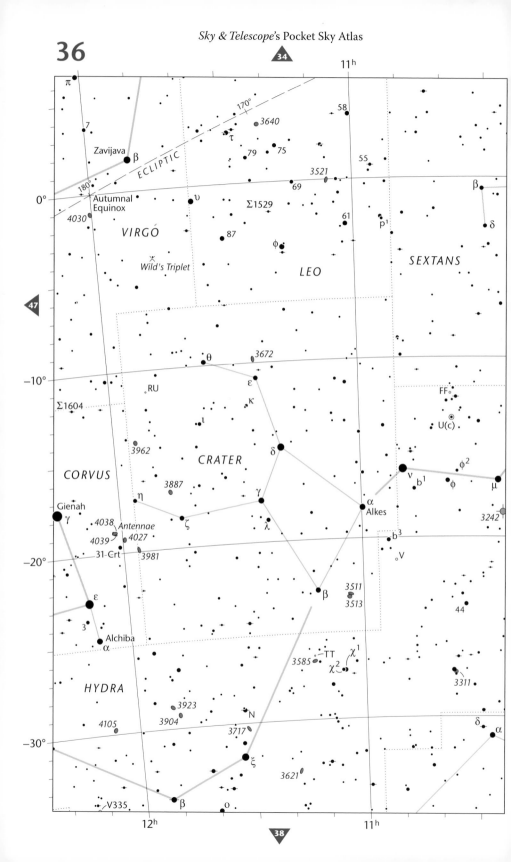

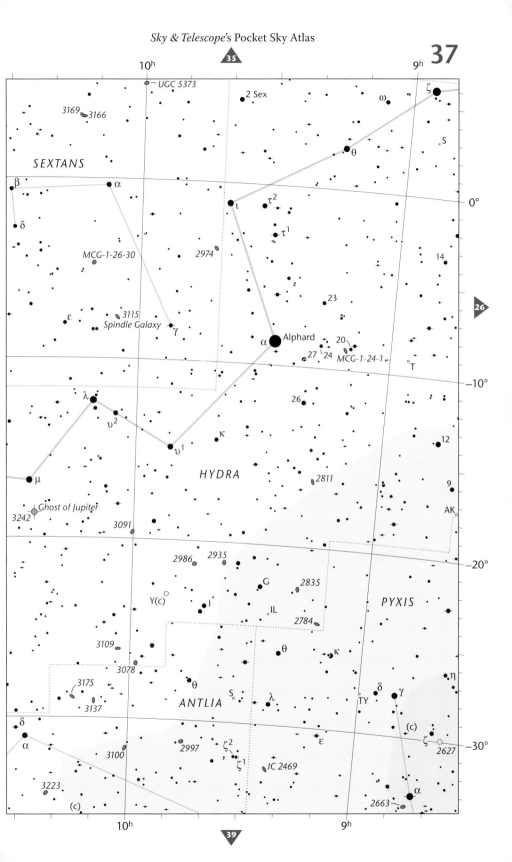

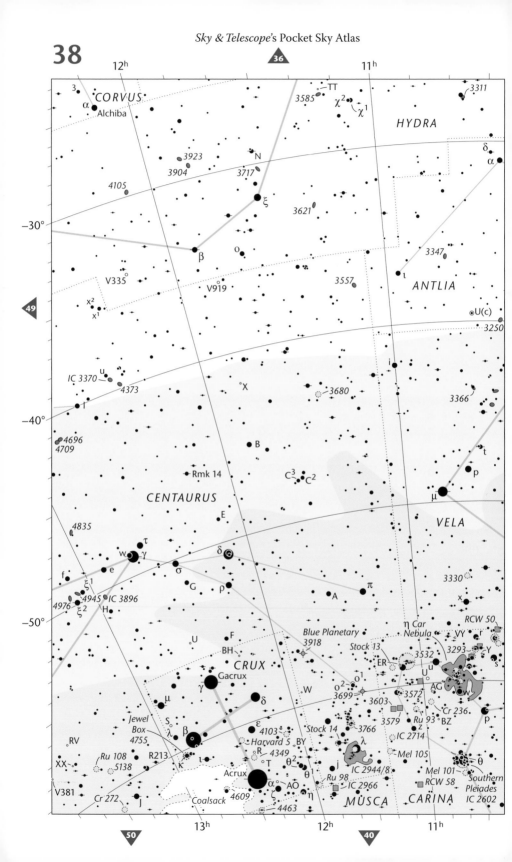

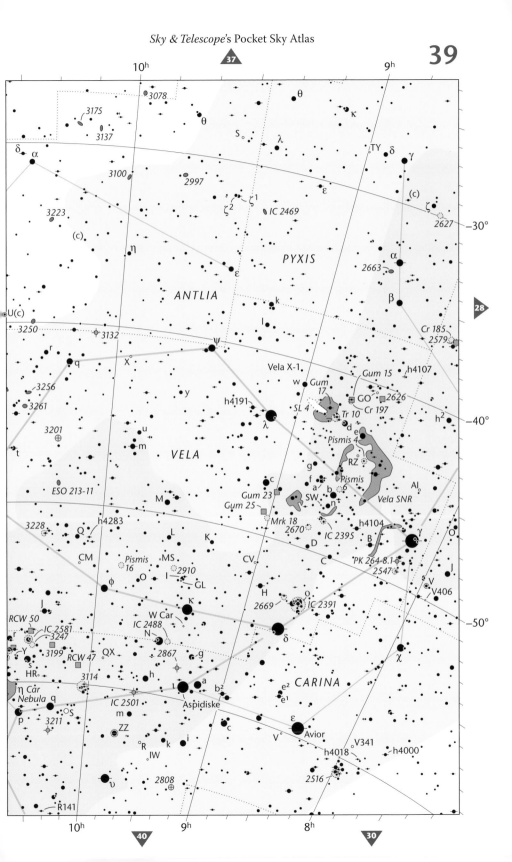

38

39

50

CENTAURUS

Blue Planetary
3918

Stock 13 · 3532

RCW 50

IC 2581

3247

VY

r

3293

3199

VELA

κ

WCar

N

IC 2488

2867

g

QX

RCW 47

3114

h

IC 2501

a

ι

b²

u

U

AG

ER

o²·o¹

3699·

3603·

3572·

w

η Car
Nebula

Cr 236

BZ

q

p

S

Aspidiske

m

c

W

3766

3579

z

3211

ZZ

R

k

i

IW

CARINA

−60°

CRUX

Stock 14

4103

λ

IC 2714

Mel 105

Southern Pleiades

IC 2602

θ

Mel 101

2808

h4164

BY

θ²

θ¹

j

IC 2944/8

Ru 98

IC 2966

RCW 58

α

β

AO

ζ

η

R141

3136

ε

T

4463

λ

μ

ω

2822

β

IC 2448

Miaplacidus

θ

ζ²

ε

ζ¹

BO

ζ

MUSCA

VOLANS

β

α

S

R

3059

κ

η

ζ

EF

Dark Doodad

ν

−70°

4833

γ

4372

π

Be 142

CHAMAELEON

R

α

θ

ι²

ι¹

Sa 156

ε

γ

RS

η

β

δ¹

δ²

3195

ζ

ι

ε

μ

θ

θ

X

η

ζ

15h

α

ε

η

U

ι

ζ

κ

ι

IC 4499

δ

κ

ω

π

16h

APUS

π²·π¹

ρ

TZ

R

ξ

MENSA

ν

U

δ

δ

γ

σ

South Celestial Pole

IC 2051

β

S

χ

17h

6438

τ

υ

18h

OCTANS

HYDRUS

μ

τ²

19h 20h 21h 22h 23h 0h 1h 2h

11h 10h

Charts 41–50

Right Ascension 12ʰ to 15ʰ

This section is highest in the night sky around these times:

Evening: **May & June**
Midnight: April
Morning: February & March

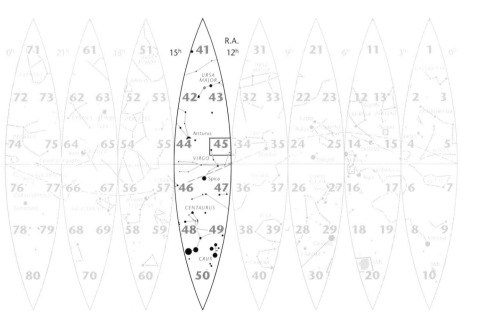

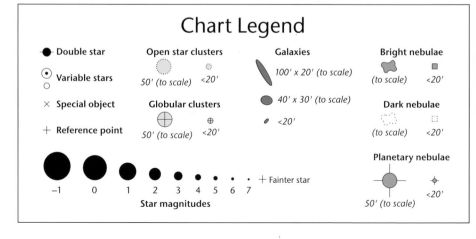

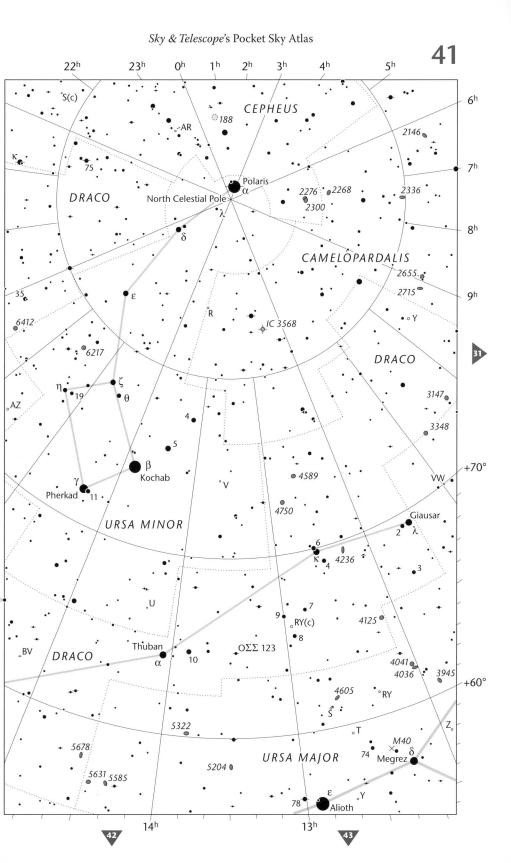

CEPHEUS

°S(c)

AR

☼ 188

2146

κ

75

DRACO

Polaris
α
North Celestial Pole

2276
2300

2268

2336

λ

δ

CAMELOPARDALIS

2655

35

ε

2715

6412

R

IC 3568

Y

6217

DRACO

31

η
19

ζ
θ

3147

AZ

4

3348

5

4589

β
Kochab

VW

+70°

γ
Pherkad

11

V

4750

Giausar

URSA MINOR

6

2

λ

κ

4236

U

4

3

7

4125

9

RY(c)

8

BV

DRACO

Thuban

OΣΣ 123

4041
4036

3945

α

10

+60°

4605

RY

Š

5322

T

Z

5678

M40

5631 5585

5204

74

×
Megrez

δ

URSA MAJOR

ε

γ

78

Alioth

22ʰ 23ʰ 0ʰ 1ʰ 2ʰ 3ʰ 4ʰ 5ʰ

6ʰ

7ʰ

8ʰ

9ʰ

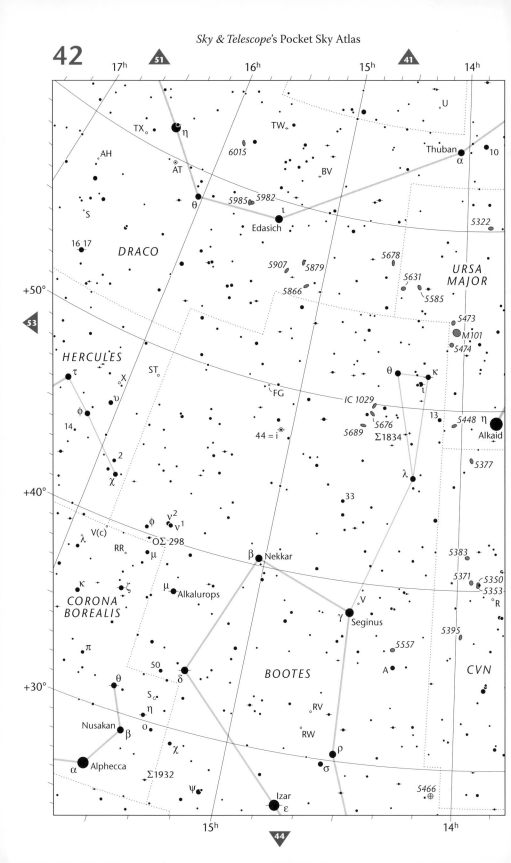

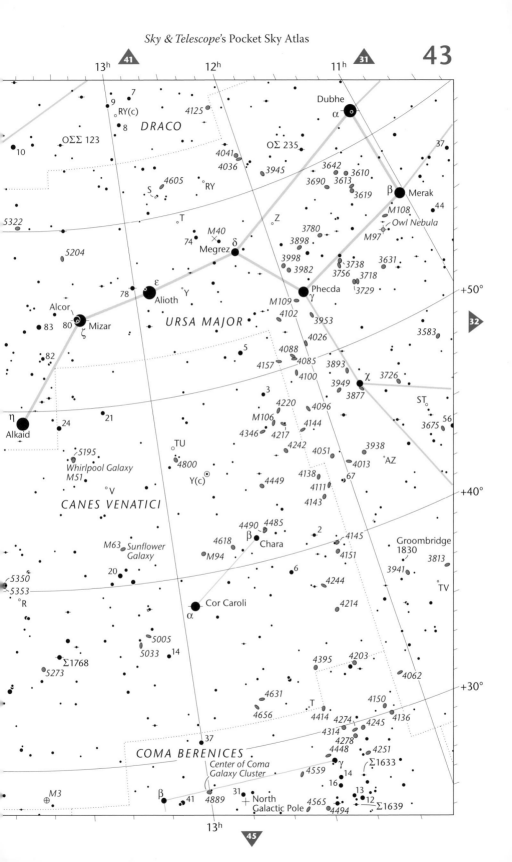

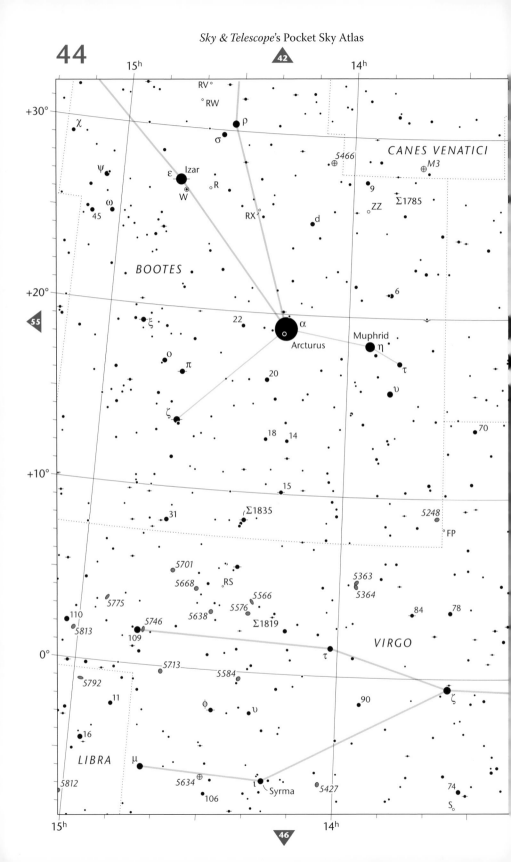

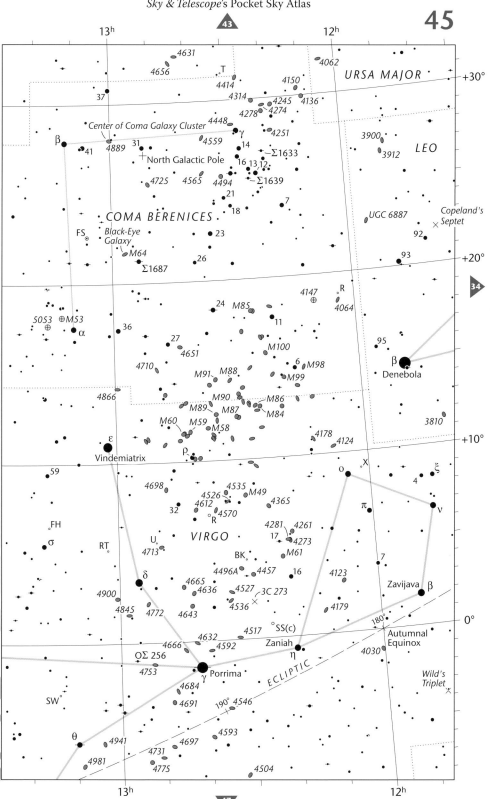

13ʰ
12ʰ

4631
4656
T
4414
4150
4062
URSA MAJOR
+30°

37
4314
4245
4136
4278 4274

Center of Coma Galaxy Cluster
4448
γ
4251
3900
LEO
β
4889
31
4559
14
3912
41
Σ1633
North Galactic Pole
16 13 12
4725
4565
4494
Σ1639
21
7
UGC 6887
Copeland's
Septet
18
92

COMA BERENICES
23
93
+20°
FS
Black-Eye
Galaxy
M64
26
Σ1687

4147
R
5053
M53
36
24
M85
4064
α
11
95
27
M100
β
4651
4710
6
M98
Denebola
4866
M91 M88
M99
M90
M86
3810
M89 M87
M84
M60 M59
M58
4178
4124
+10°
ε
ρ
X
Vindemiatrix
o
ξ
59
4
4698
4535
π
ν
4526
M49
FH
4612
4365
7
σ
32
R 4570
4281 4261
RT
U
VIRGO
17
4273
Zavijava
β
4713
BK
M61
4123
δ
4496A
4457
16
4900
4665
4527
3C 273
4179
4845
4636
4536
4772
4643
SS(c)
4517
Autumnal
Equinox
0°
4632
4592
ΟΣ 256
4666
Zaniah
4030
4753
γ
Porrima
η
Wild's
Triplet
SW
4684
ECLIPTIC
4691
190°
4546
θ
4941
4593
4981
4731
4697
4775
4504

34

15ʰ

14ʰ

5668   RS   5566   84   78
5638   5576   Σ1819
5775
5838   5746
5813   110   109   5584   τ
5850 5846   90   ζ
5792   5713
0°   φ   ν   *VIRGO*
11
16   μ   5634 ⊕ ι   5427   74
Syrma   S
106   82
Y   5812   95
δ   κ   76
β Zubeneschamali   210°
−10°   17 ξ²
FY   ξ¹   λ   ECLIPTIC
μ   220°
α¹   89 87   5247
o   ν   α² Zubenelgenubi   DL
230°   5728
LIBRA   47
−20°   ι
S   5897   HN 28
RS   12
5903   54   5694   51   50   π   *HYDRA*
σ   56   W
(c)   59   58   52   Seashell Galaxy
RU   M83
5253
−30°   2   *CENTAURUS*   h   IC 4296
1   T
LUPUS   5824   5419   2   ESO 383-87
c² c¹   y   z

57 ◀

15ʰ   48 ▽   14ʰ

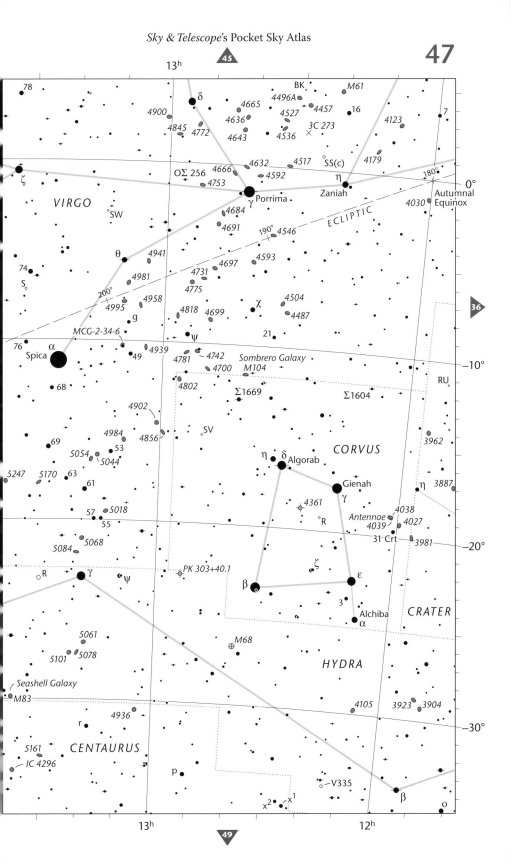
13ʰ

78

δ

4900
4845   4772

4665
4636
4643

BK
4496A
4527
4536

M61
4457
16

3C 273
×

4123

7

ζ

VIRGO

°SW

OΣ 256   4666
4753

4632
4592

4517

4684

SS(c)

η
Zaniah

180°

0°
Autumnal
Equinox

4030

ECLIPTIC

4691
190°   4546

θ

4941

4981

4731
4775

4697

4593

4504

χ

4487

74

S°

200°
4995

4958

g

4818   4699

21

MCG-2-34-6

ψ

76

α
Spica

49   4939

4781   4742
4700

4802

Sombrero Galaxy
M104

-10°

RU°

68

Σ1669

Σ1604

4902

SV°

3962

4984

4856

η   δ Algorab

CORVUS

69

53

5054

5044

5247   5170   63

61

Gienah
γ

η   3887

4361

°R

Antennae
4039   4038
4027

57   5018

55

5084   5068

31 Crt   3981

-20°

R°

γ

ψ

PK 303+40.1

ζ

β

ε

3

Alchiba
α

CRATER

5061

M68
⊕

HYDRA

5101   5078

Seashell Galaxy
M83

4105   3923   3904

r   4936

-30°

5161
IC 4296

CENTAURUS

p

V335

β

o

x²   x¹

13ʰ

49

12ʰ

15ʰ

46

14ʰ

47

12
54
56
σ
59
58
50
π
W

5694
51

RU
Seashell Galaxy
M83

HYDRA
52

LIBRA

5253

h
1
T
5161
IC 4296

2
5419
2

CENTAURUS
ESO 383-87

−30°
2
1
y
z

5824

Menkent
θ

LUPUS
c² c¹
V1002
ψ
5367
d

φ¹
φ²
5408
ν
χ
μ

k
φ
5128

υ
5786
η
υ¹
ESO 270-17

δ GG
κ β
o
5483
5530
IC 4444
IC 4406
5643
υ²

−40°
γ
ω
e
λ
h4690
τ¹
τ²
ι
IC 4402
ζ
ω Cen 5139

ε
5882
π
ESO 221-26 5460
5266
V744

g d
ESO 274-1
μ
α
ρ
5307
5286
K
M

ν¹
ν² κ
σ
ε
Q

R
Hogg 18 b
c
V716
RV

η
5927
5946
ζ
5822
5749
v
V381 XX
Ru 108
5138

−50°
5823
5662
V
V412
V760
R

RCW 103
T
V737
β
Rigil Kent
β
Cr 272
5316
Tr 21

RCW 102
6067
NORMA
Cr 292
PK 322-2.1
CIRCINUS
β Pismis 20
α
5606
5617 Tr 22
Lynga 2
V766
5281

κ
Cr 299 ζ
ι¹
γ
δ
θ
Proxima
Centauri
Ced 122
m

6087
ι²
S
R
vdBH 65a
Sa 172

ESO 137-34
h4813
6025

16ʰ

15ʰ

14ʰ

60

50

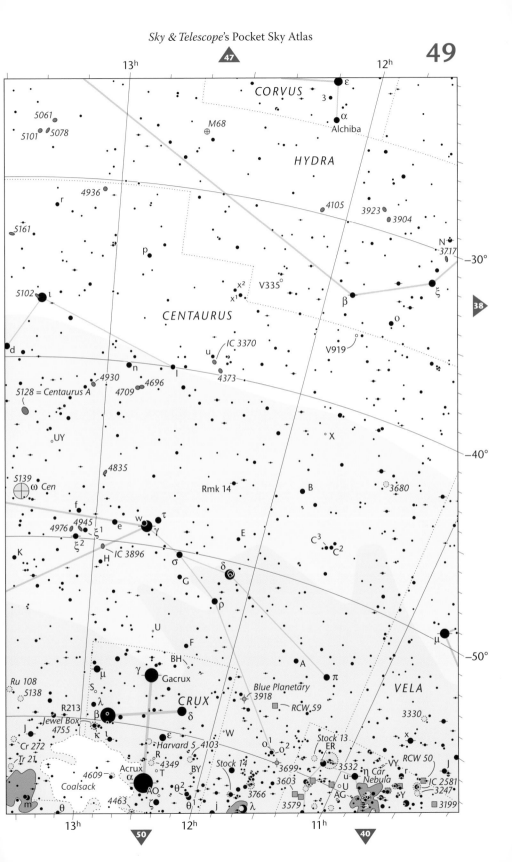

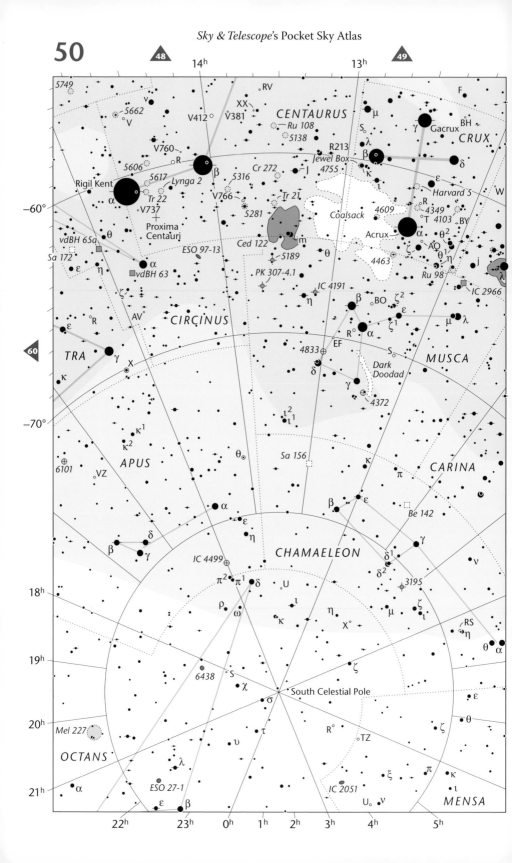

Charts 51–60

Right Ascension 15^h to 18^h

This section is highest in the night sky
around these times:

Evening: **July**

Midnight: May & June

Morning: April

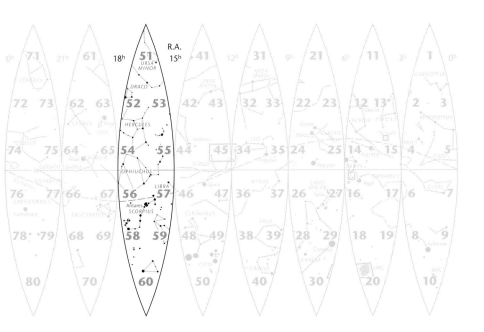

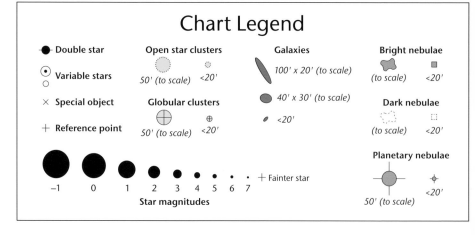

Chart Legend

Double star

Variable stars

× **Special object**

+ **Reference point**

Open star clusters

50' (to scale)   <20'

Globular clusters

50' (to scale)   <20'

Galaxies

100' x 20' (to scale)

40' x 30' (to scale)

<20'

Bright nebulae

(to scale)   <20'

Dark nebulae

(to scale)   <20'

Planetary nebulae

<20'

50' (to scale)

−1   0   1   2   3   4   5   6   7   + Fainter star

Star magnitudes

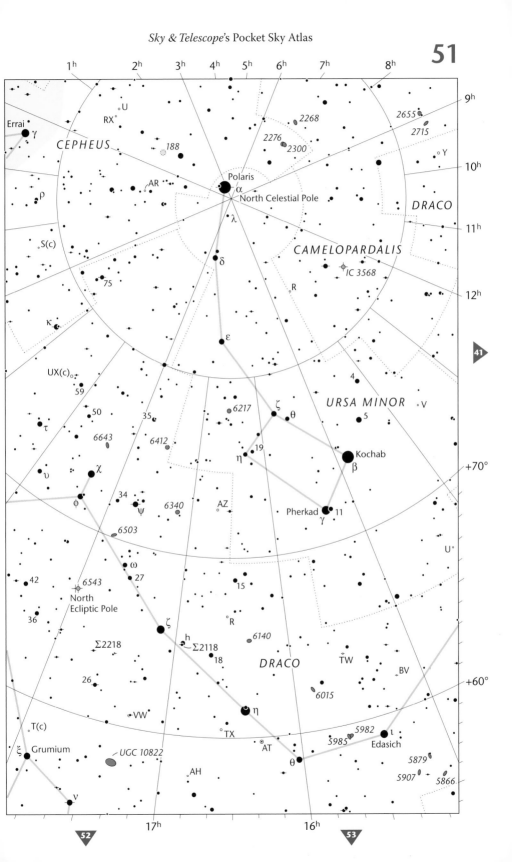

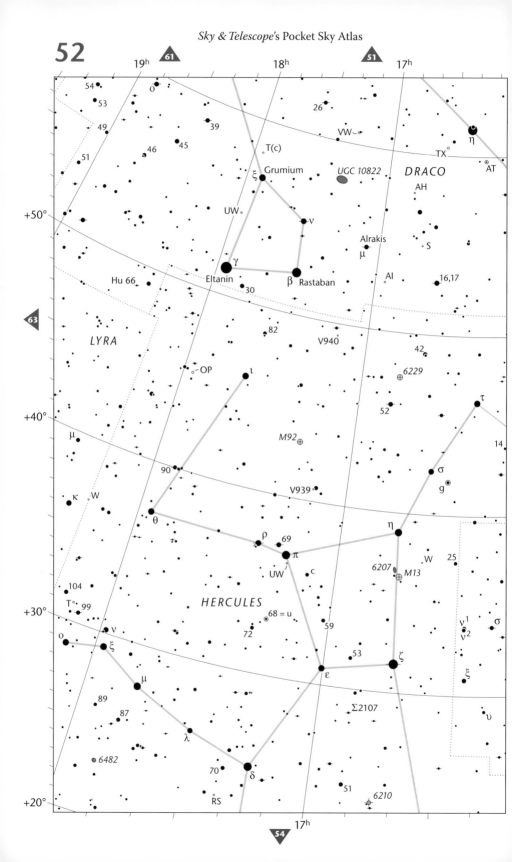

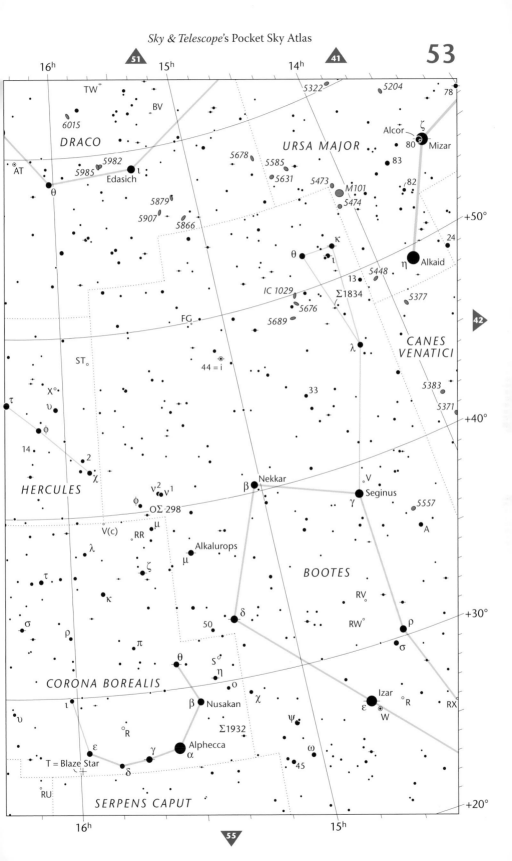

51 16ʰ 15ʰ 14ʰ 41

TW

BV

5322 5204 78

6015

DRACO URSA MAJOR Alcor ζ
80 Mizar

5982 5678 83
5985 5585
Edasich 5631 5473 82
AT θ ι 5473 M101
5879 5474 +50°

5907 κ 24

5866 θ ι η Alkaid

IC 1029 13 5448
Σ1834 5377

FG 5676 CANES
5689 VENATICI

λ 5383
44 = i 5371

ST 33 +40°

X
τ υ
φ 5557
14 V
2 Nekkar Seginus
χ β γ

HERCULES ν²ν¹ A

φ OΣ 298
μ BOOTES +30°
V(c) RR
λ μ Alkalurops RV
τ ζ μ RW
κ ρ
σ δ σ +30°
ρ 50
π θ S Izar R
CORONA BOREALIS η ε RX
o χ W
β Nusakan ψ
ι Σ1932
υ R ω 45
ε γ Alphecca +20°
T = Blaze Star δ α
RU SERPENS CAPUT

42

18ʰ
17ʰ

104
T
99
+30°
ν
ξ
o
μ
89
87
λ
70
δ
RS
6482
98
95
96
+20°
65
93
Z

68 = u
72
59
53
ζ
ε
Σ2107

HERCULES

51
6210
SY

Kornephoros β
s
54
γ
U
Σ2052

Rasalgethi
α
60
S
ω
29

Rasalhague
α
e
37
ι
κ
43
47
45

+10°

6384
IC 4665
vdB 111
β Cebalrai
σ
Barnard's Star
γ
6426
Cr 350
V2113
Z U
0°
41
OPHIUCHUS
V533
M14

λ Marfik
σ

M12
6118
δ
Yed Posterior ε
M10
6366
30
23

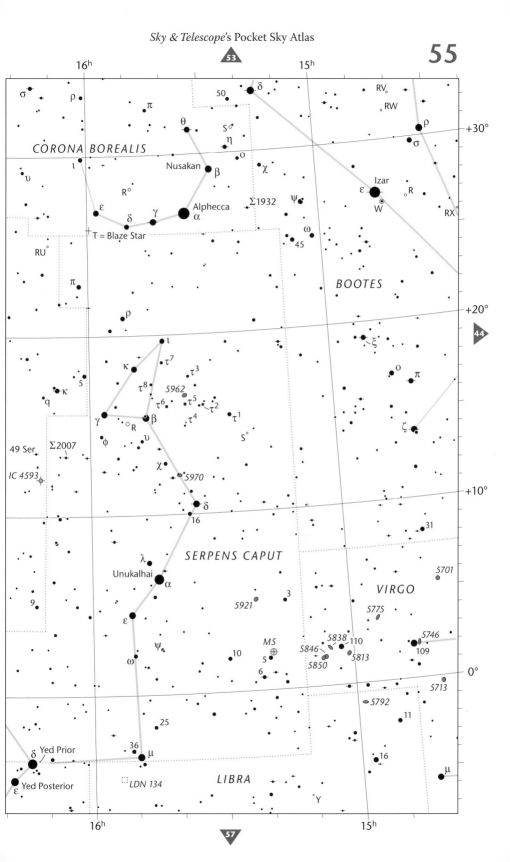

44

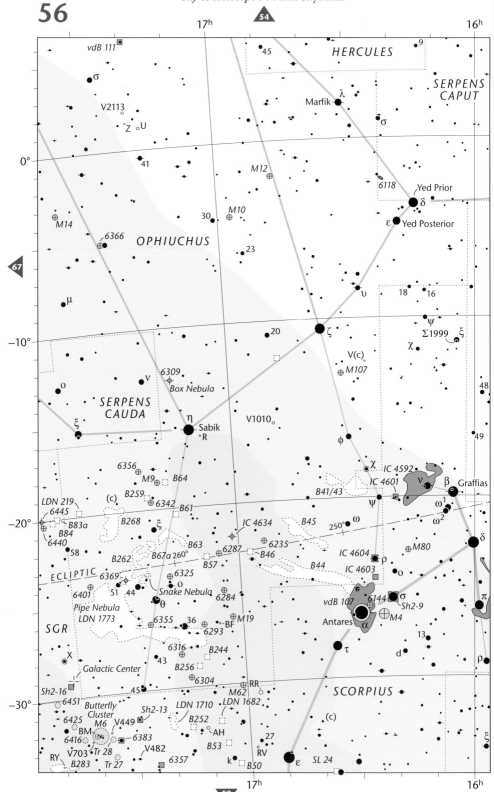

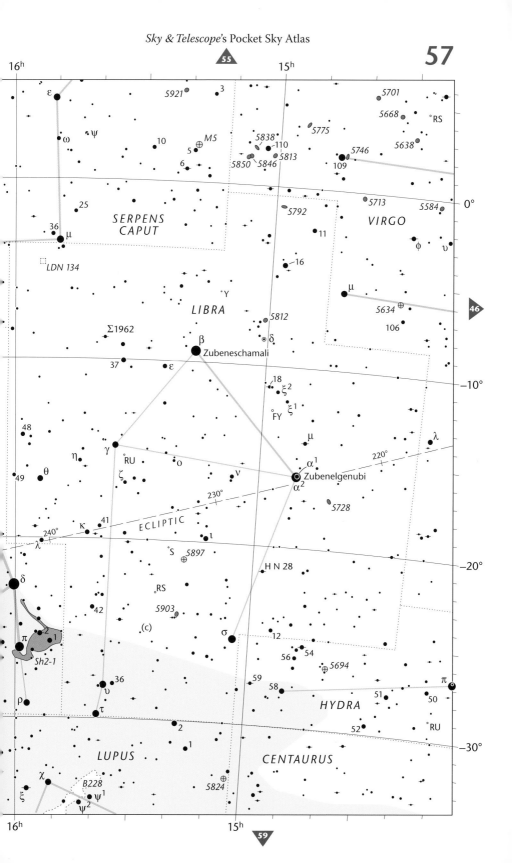

16ʰ

15ʰ

ε

5921   3

5701
5668   °RS

ω   ψ

10   M5   5838   110
5   5850 5846 5813   5775   5638

6   5746
109

25

5713   5584   0°

SERPENS
CAPUT

5792

VIRGO

36
μ   11   φ   υ

LDN 134   16

°Υ   μ

LIBRA   5634
106

Σ1962   5812

β   δ
Zubeneschamali

37   ε   18
ξ²

ξ¹
°FY

48   μ   λ

η   γ   220°

49   θ   °RU   o   α¹
ζ   ν   Zubenelgenubi
α²

230°   5728

κ   41
240°   ECLIPTIC
λ   ι

°S   5897   H N 28   −20°

°RS

42   σ
5903   12

(c)   56   54
5694

π   2 1   59   π
Sh2-1   36   58   51
ρ   υ   50

τ   2   52   °RU   −30°

LUPUS   1   CENTAURUS

HYDRA

χ
ξ   B228   5824
ψ¹
ψ²

16ʰ   15ʰ

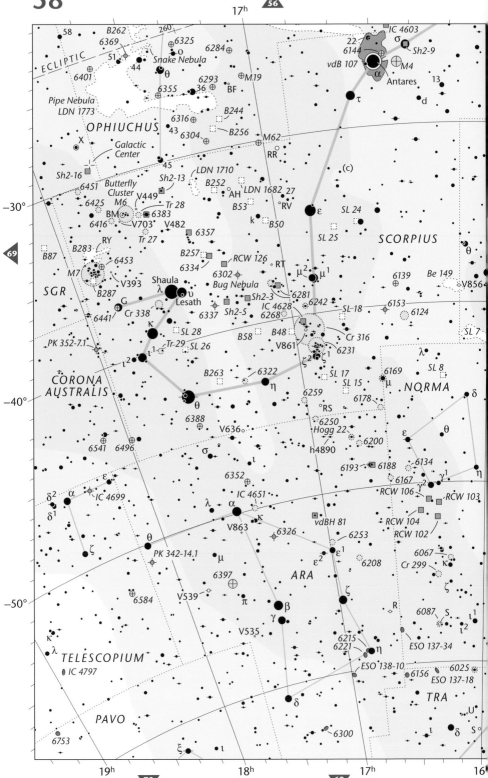

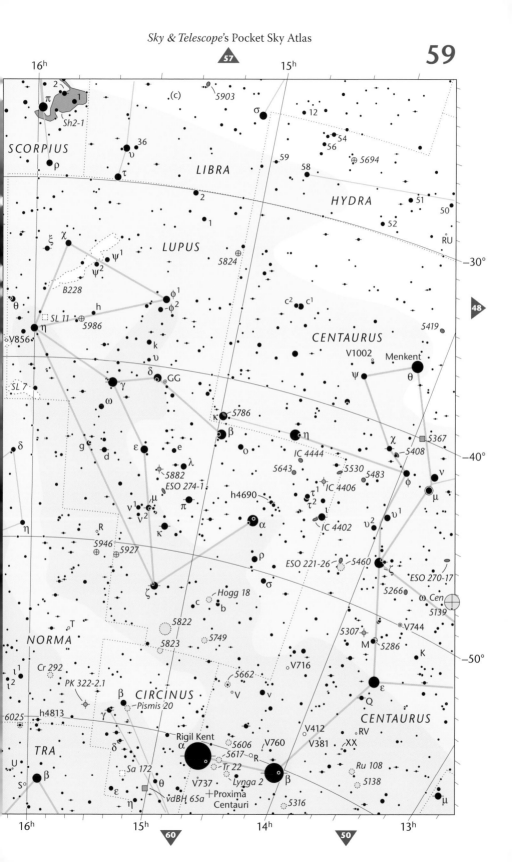

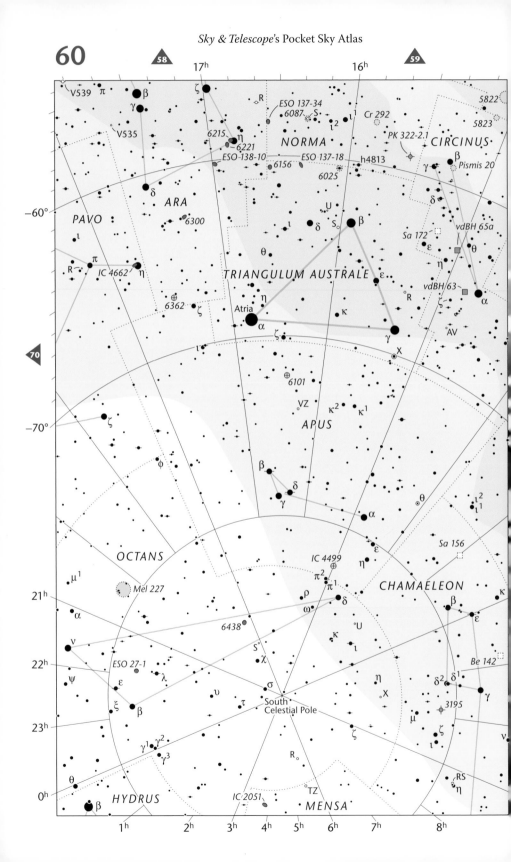

17ʰ

16ʰ

V539 π β
γ
V535
δ
ARA
PAVO
6300
ι
π
R
η
IC 4662
6362 ζ

R
6087 S
ESO 137-34
6215
η
6221
ESO-138-10
6156
ESO 137-18
6025
h4813
NORMA
Cr 292
PK 322-2.1
CIRCINUS
β
γ Pismis 20
δ
Sa 172
ε θ
vdBH 65a
5822
5823

ι¹
ι²

U
S
δ
ι
θ
β
ε
R
η
vdBH 63
α
ζ
AV

TRIANGULUM AUSTRALE
η
Atria
α
ζ
ι
κ
γ
X

6101
VZ
κ² κ¹
APUS

−60°

−70°

70

ζ
φ

β
γ δ
α
θ
ι¹
ι²

ε
η
Sa 156
θ

OCTANS
μ¹
Mel 227

IC 4499
π²
π¹
ρ
ω
δ

CHAMAELEON
β
ε κ

21ʰ
α
6438
S
χ
κ
ι
U

η
X
Be 142
δ² δ¹
γ

22ʰ
ν
ψ
ESO 27-1
ε
λ
ξ β
υ
τ
σ
South
Celestial Pole

μ
3195
ζ
ι

23ʰ
γ¹ γ²
γ³
ζ
RS
η
ν

0ʰ
θ
HYDRUS
β
IC 2051
R
TZ
MENSA

1ʰ   2ʰ   3ʰ   4ʰ   5ʰ   6ʰ   7ʰ   8ʰ

Charts 61–70

Right Ascension 18ʰ to 21ʰ

This section is highest in the night sky
around these times:

Evening: August & September

Midnight: July

Morning: May & June

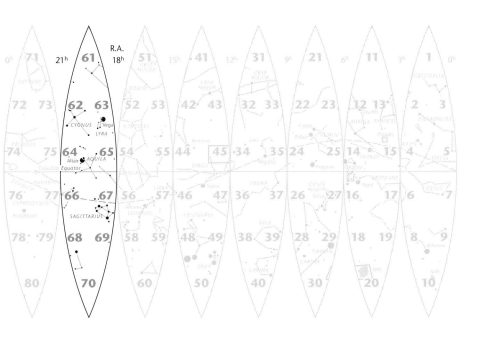

Chart Legend

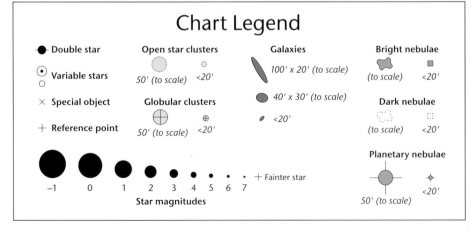

Double star

Variable stars

× **Special object**

+ **Reference point**

Open star clusters
50' (to scale)   <20'

Globular clusters
50' (to scale)   <20'

Galaxies
100' x 20' (to scale)
40' x 30' (to scale)
<20'

Bright nebulae
(to scale)   <20'

Dark nebulae
(to scale)   <20'

Planetary nebulae
50' (to scale)   <20'

−1   0   1   2   3   4   5   6   7   + Fainter star
Star magnitudes

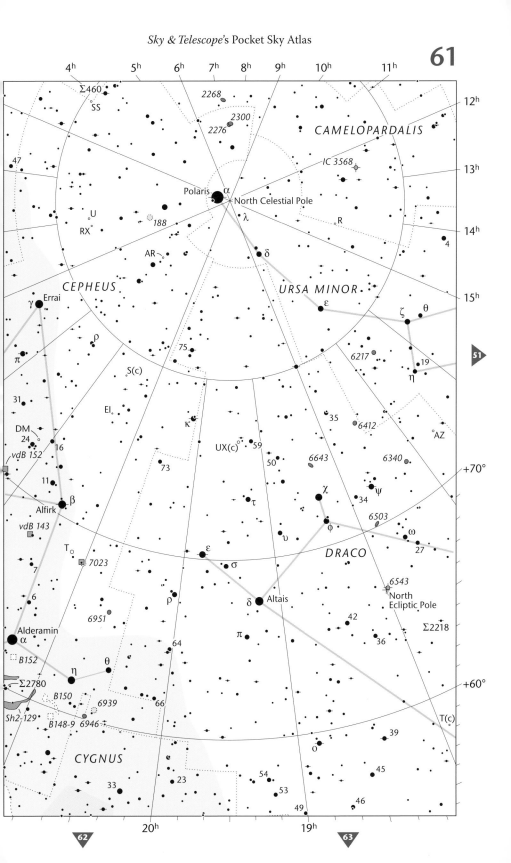

4ʰ 5ʰ 6ʰ 7ʰ 8ʰ 9ʰ 10ʰ 11ʰ

12ʰ

13ʰ

14ʰ

15ʰ

+70°

+60°

19ʰ

20ʰ

Σ460
SS
2268
2300
2276
CAMELOPARDALIS
IC 3568
47
Polaris
α
North Celestial Pole
4
U
RX
188
λ
R
AR
δ
CEPHEUS
URSA MINOR
ε
ζ θ
γ Errai
75
19
η
ρ
6217
π
S(c)
31
EI
κ
35
6412
AZ
DM
24
UX(c) 59
6643
6340
vdB 152
16
50
11
73
χ
ψ
β
τ 34
Alfirk
υ φ
6503
vdB 143
ω
27
T
ε
DRACO
7023
σ
7
ρ
6
δ Altais
6543
North Ecliptic Pole
Alderamin
6951
42
Σ2218
α
π
36
B152
θ
η
39
Σ2780
o
B150
6939
45
Sh2-129
66
B148-9 6946
CYGNUS
54
33 23
53
46
49
T(c)

51

62 63

22ʰ 21ʰ 20ʰ

71 61

CEPHEUS

Σ2840
IC 1396
vdB 140
6946 6939
B148-9
CN
23
β
7243
RU
7128
B364
FZ
+50°
7008
33
V1143
π¹
B164
Le Gentil 3
20
π²
7086
B362
ψ
Blinking
Planetary
Cocoon
Nebula
IC 5146
vdB 145
26 16 θ ι
CH
R
7209
B168
M39
51
7082
LDN 970
IC 1369
ω²
Z
RT
7062
71
LDN 970
ω¹
U(c)
o²
6811
V1339
W ρ
63 59
(c)
Sh2-115
o¹ 30
AF
75
7048
IC 5076
O O
68
7039
60
55
Deneb
Sh2-112
OΣ 406
6
North America Nebula
7000
α
δ
77
74
V1070
57
ξ
IC 5070
B346
AX(c)
6866
RR
7044
Pelican Nebula
7027
IC 5068
14
(c)
72
σ
ν
Northern
Coalsack
6914
V973
6819
70
τ
61
6910 Cr 419
IC 1311
6888 19
7063
RW
V367
γ Sadr
IC 1318
Crescent
Nebula
Piazzi's Flying Star
M29
RS(c)
6888 19
PK 80-6.1
Egg Nebula
vdB 133
Berk 86
P
Basel 6
Ced 174
22
υ
λ
IC 4996
25
15
X
Sh2-104
27 OΣ 394
Y
47
6883
6871
4
T
35
Biur 2
Sh2-101
η
ε
Σ2576
8
ζ
6992
V449
17
(c)
6995
39
vdB 128
χ
PK 64+5.1
Veil Nebula
52
φ
9
6960
CYGNUS
41
IC 4954-5
6834
SU
Footprint Nebula = M1-92
32 T
6940
21
Albireo
31
23
15
SV
Sh2-90 6813
β
PK 72-17.1
30
19
16
Sh2-88
Stock 1
6800
6882
R
6885
Stock 1
6823
FI
28
24
α
VULPECULA
33
17
13
6830 6820
vdB 126
29
22
Dumbbell Nebula
6823
M27 12
DELPHINUS
6905
θ
6886 X

+40°
+30°
+20°

73

21ʰ 20ʰ

64

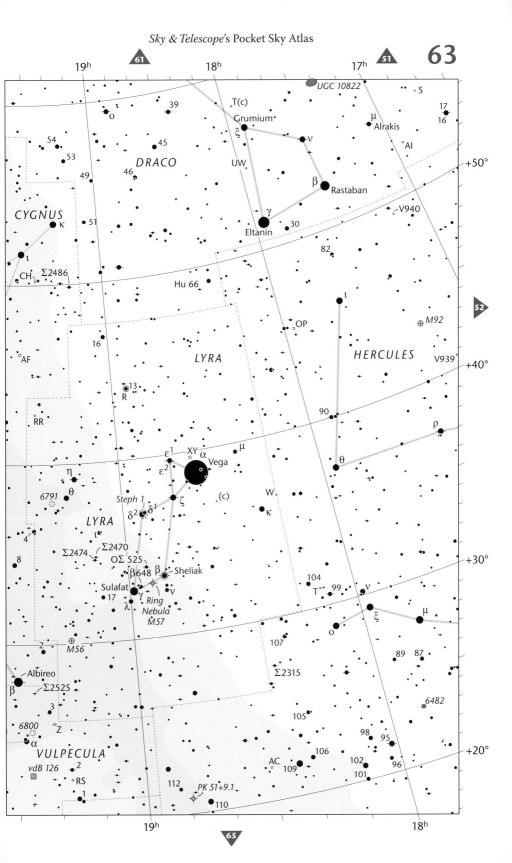

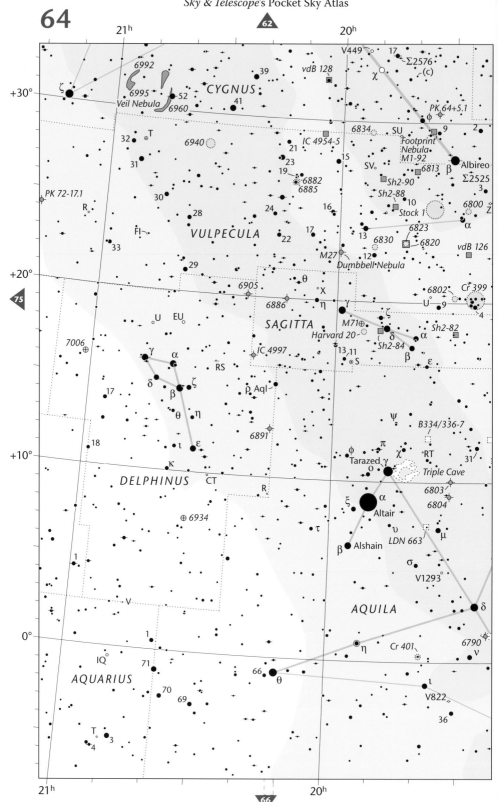

19ʰ

18ʰ

M57 β Sheliak
Sulafat *Ring Nebula*
17 γ β648 ν
λ

104
T 99
ν
ξ
o
μ

+30°

M56

2

LYRA
Σ2315
107

89 87

Σ2525
3

6800 Z

α

VULPECULA

vdB 126 2
RS
1

105

6482

AC 106 98
109 95
96

+20°

112
PK 51+9.1
110

102
101

54

Cr 399
Brocchi's
Cluster
4 U

HERCULES

IQ

93

111

Z

FF

OΣ 358

ε

RX

ζ 11

LDN 684

Σ2404

+10°

ω
31
LDN 673
18

6709

PK 38+12.1

72

AQUILA

X

OPHIUCHUS 71

R

6781
Σ2497

6633
6572

IC 4665

β
Cebalrai
66
Barnard's
Star

6756

Σ2375
IC 4756

73

6755

RY
74

70 67
68

γ
6426

δ

θ Alya

Cr 350

21
Sh2-72

4

23 6760
6790 TT
ν

vdB 123

59 = d

0°

6535

*SERPENS
CAUDA*

B138

V533

27
6741

LDN 564

LDN 582
60
Sh2-64

ζ

B139

LDN 557

η

LDN 581 14
15
LDN 567

B111
β
R

SCUTUM

f
6751
B134 V(c) 12

B119a η
M11

B103

IC 1276

19ʰ

18ʰ

20h

64

V

21
6790
23 6760
η
Cr 401
ν
B138
1
66
θ
ι
27
IQ
V822
71
36
70
69
AQUILA
AQUARIUS
f
T
3
κ
U
W
4
Pal 11
20

77
μ
ε Albali
51
6814
37
(c)
−10°
α²
ν
α¹
Little Gem
M72
ξ
6818
M73
Algedi
V505
(c)
τ
β² Dabih
Barnard's
υ
β¹
61
Galaxy
55
54
T
310°
ρ
6822
ρ¹
υ
ο
π
(c)
ρ²
σ
43
R
300°
56
−20°
η
RT(c)
ECLIPTIC
M75
290°
CAPRICORNUS
χ³
6907
52
χ¹
ψ
24
60 ω
SAGITTARIUS
ψ
59
ω
62
V1943
RR
T
M55
−30°
h5218
δ
6925
RY
γ
β
α
θ²
MICROSCOPIUM
θ¹

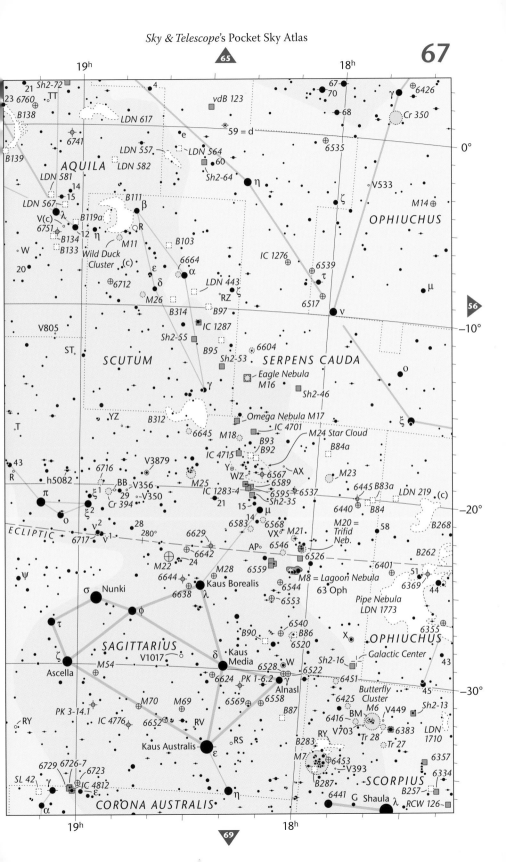

19ʰ   18ʰ

21 Sh2-72
23 6760   TT
B138

0°

LDN 617

vdB 123
67
70   γ   6426
68   Cr 350

V533

6741
e   59 = d
6535

LDN 557   LDN 564
LDN 582   60
Sh2-64   η   ζ   M14

AQUILA
LDN 581
14
15
B111   β
LDN 567
V(c) λ   B119a
6751
12 η   R   B103
B134
W   B133   M11
Wild Duck   ε   B103
20   Cluster (c)   δ   6664   α   LDN 443
6712   δ   RZ ζ
M26   B314   B97

OPHIUCHUS

ζ

μ
6539
τ
6517
ν

IC 1276

V805
V805   –10°
ST
SCUTUM
IC 1287
Sh2-55   B95
Sh2-53   6604
SERPENS CAUDA
o
Eagle Nebula
M16
Sh2-46
ξ

YZ   B312
Omega Nebula M17
6645   M18   IC 4701
M24 Star Cloud
B93
IC 4715   B92   B84a
T   6716   AX
V3879   Y   M23
R   h5082   BB V356   WZ   6567   6445 B83a   LDN 219
π   ξ¹   29   M25   6589
ξ² Cr 394 V350   IC 1283-4   6595   6537   6440   B84
o   28   21 15 μ   Sh2-35
v²   280°   14   6568   M20 =
ECLIPTIC   v¹   6629   6583   VX M21   Trifid
6717   6642   AP   6546   Neb.   B268
24   6526   B262
M22   M28   6559   M8 = Lagoon Nebula   6401
6644   Kaus Borealis   63 Oph   51
ψ   σ Nunki   6638 λ   6544   Pipe Nebula   6369 44
6553   LDN 1773
6540   B86   6355
SAGITTARIUS   V1017   δ   B90   6520   X   OPHIUCHUS
τ   φ   Kaus   6528   W   Sh2-16   Galactic Center   43
ζ   Media   6522   Sh2-16
Ascella M54   6624 PK 1-6.2 γ   6451   45
M70 M69   Alnasl   6416   Butterfly
PK 3-14.1 IC 4776   6569   6558   Cluster   Sh2-13
6652   RV   B87   BM M6 V449
Kaus Australis   RS   RY V703 Tr 28   6383   LDN
ε   B283   6453   Tr 27   1710
6729 6726-7   M7 V393   6357
SL 42 γ 6723   B287   6334
α IC 4812   SCORPIUS   B257
CORONA AUSTRALIS   η   6441 G Shaula λ RCW 126

56

–20°

–30°

19ʰ   18ʰ

69

–20°

ξ

μ

o

43

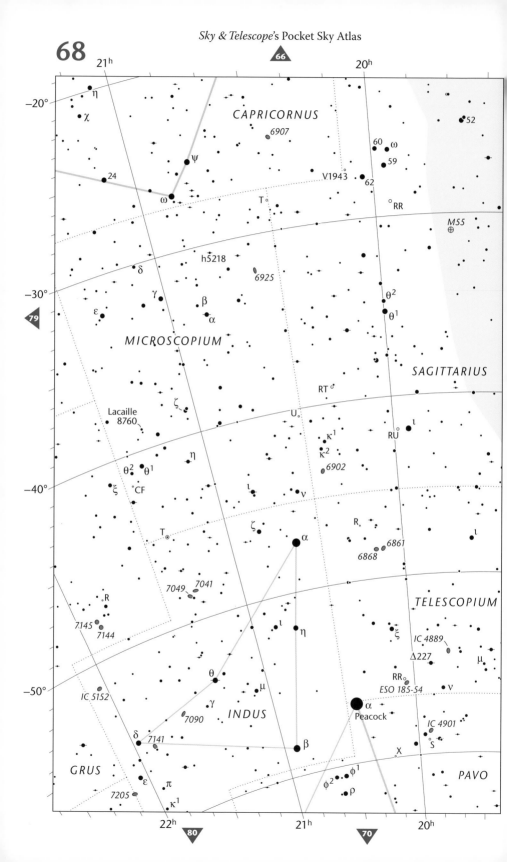

21ʰ

20ʰ

66

−20°

η
χ

CAPRICORNUS

6907

52

ψ

60 ω
59

24

V1943
ω

62

RR

T

M55

h5218

δ

6925

θ²
θ¹

−30°

γ

β

ε

α

MICROSCOPIUM

SAGITTARIUS

79

RT

ζ

U

Lacaille
8760

κ¹

RU ι

η

κ²

6902

θ² θ¹

ξ   CF

ι

ν

−40°

R

ι

T

ζ

6861

α

6868

7049 7041

TELESCOPIUM

R

ι
η

ξ

7145
7144

IC 4889

θ

μ

Δ227

μ

IC 5152

γ

−50°

RR
ESO 185-54

ν

7090

α
Peacock

IC 4901

δ

7141

S

INDUS

χ

GRUS

ε

π

PAVO

φ² φ¹

7205

κ¹

ρ

β

22ʰ

21ʰ

20ʰ

80

70

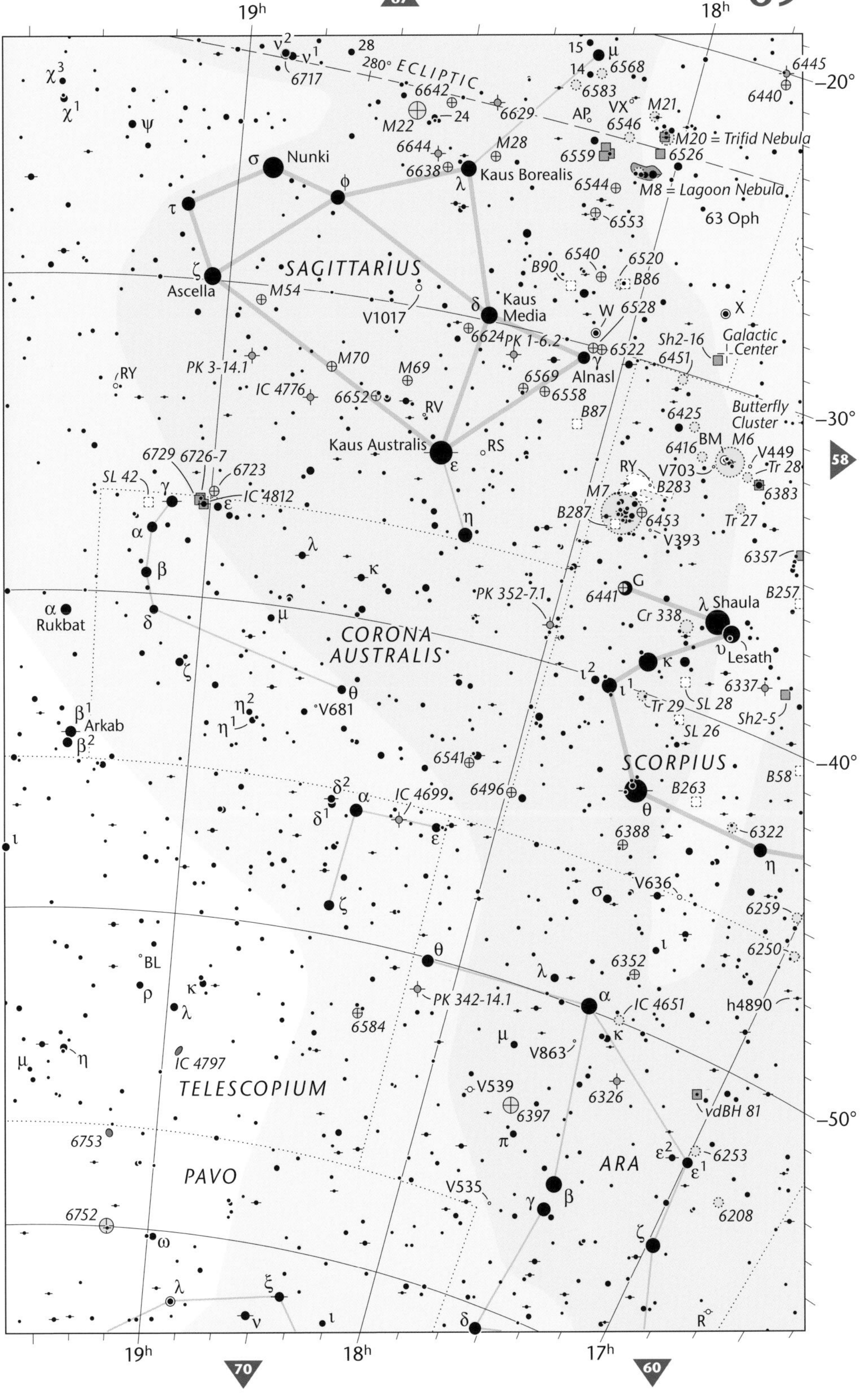
67
19h
18h
ECLIPTIC
280°
SAGITTARIUS
Nunki
Ascella
Kaus Borealis
Kaus Media
Kaus Australis
Alnasl
M8 = Lagoon Nebula
M20 = Trifid Nebula
M21
M22
M28
M54
M69
M70
M6
M7
Galactic Center
Butterfly Cluster
63 Oph
CORONA AUSTRALIS
SCORPIUS
Shaula
Lesath
Rukbat
Arkab
TELESCOPIUM
ARA
PAVO
IC 4812
IC 4776
IC 4699
IC 4651
IC 4797
PK 3-14.1
PK 1-6.2
PK 352-7.1
PK 342-14.1
–20°
–30°
–40°
–50°
58
17h
70
60

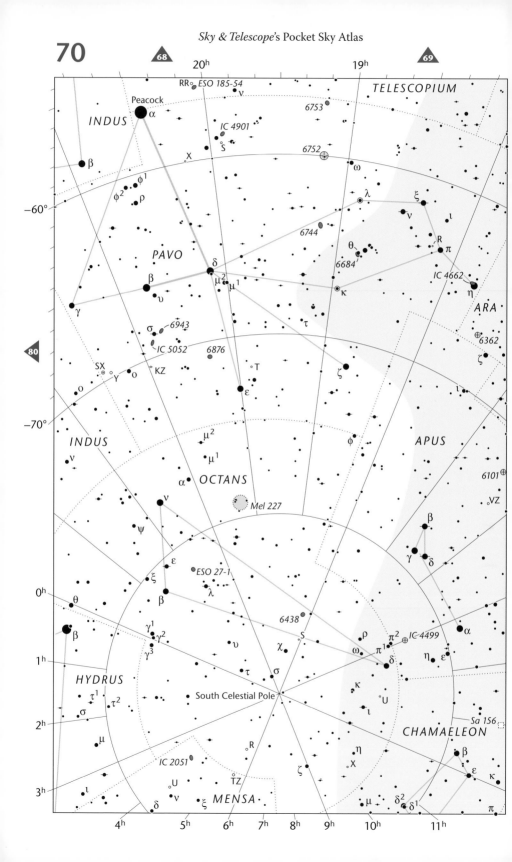

Charts 71–80

Right Ascension 21ʰ to 0ʰ

This section is highest in the night sky
around these times:

Evening: **October**

Midnight: August & September

Morning: July

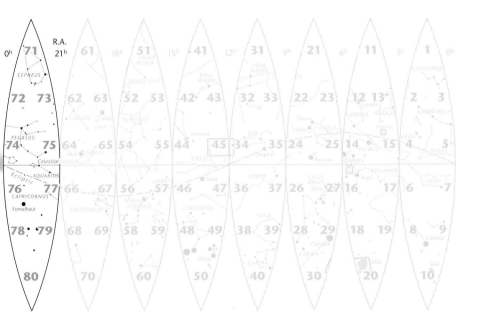

Chart Legend

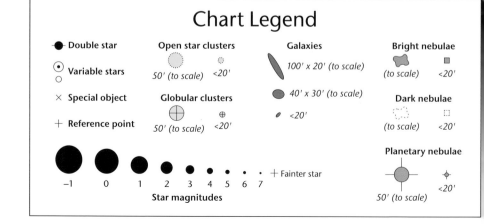

Double star

Variable stars

× **Special object**

+ **Reference point**

Open star clusters

50' (to scale)   <20'

Globular clusters

50' (to scale)   <20'

Galaxies

100' x 20' (to scale)

40' x 30' (to scale)

<20'

Bright nebulae

(to scale)   <20'

Dark nebulae

(to scale)   <20'

Planetary nebulae

<20'

50' (to scale)

−1   0   1   2   3   4   5   6   7   + Fainter star

Star magnitudes

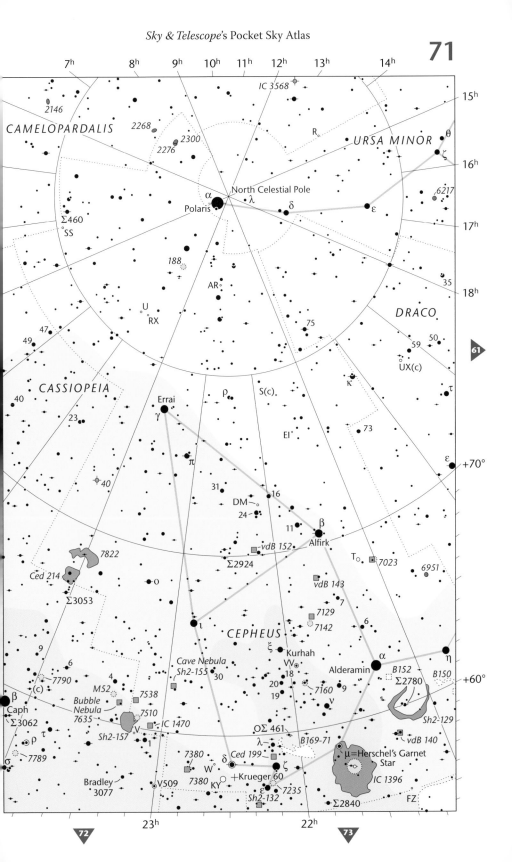

1 0h **71** 23h

CASSIOPEIA   CEPHEUS

M103 δ   *381*
χ Ruchbah   *IC 59*   κ
IC 63   *225*
Sh2-188   *457 436*   *136*   12
φ V465   *129*   9
υ²   6   *Cave*
υ¹   *7790*   M52 4   *Nebula*   30
η   *IC 10*   *7635*   7538   *Sh2-155*
θ   TV   β   *Bubble*   7510
μ   *281*   Caph   *Nebula*   *IC 1470*
α   *vdB 1*   Σ3062   *Sh2-157*   V   *7380*   δ
Schedar   ρ   τ   1   *Krueger 60* W
T   7789   σ   *Bradley 3077*   KY
ζ   λ   V509   *Sh2-132*
ε

+50°   ν   ξ   TU   R   SV   7296
278   (c)   β
RV   o *185*   18   9
π   *147*   α
7686   7   3   4
22   Z   8   5   LACERTA
ψ   11   2
Andromeda   26   λ   4
Galaxy
M31   4   15   EV   11
M110   κ   6
M32   ι   13   6
+40°   206   7662 Blue Snowball   2   12   S
32   o   10
R   θ   7640
ρ   ANDROMEDA   14   7331
σ   Stephan's Quintet ×
Σ3050   PEGASUS
+30°   V343   72   64
Alpheratz   78   PV
α   85   7457   η
Matar   o
ψ   7741   β   32
Scheat

0h   23h

74

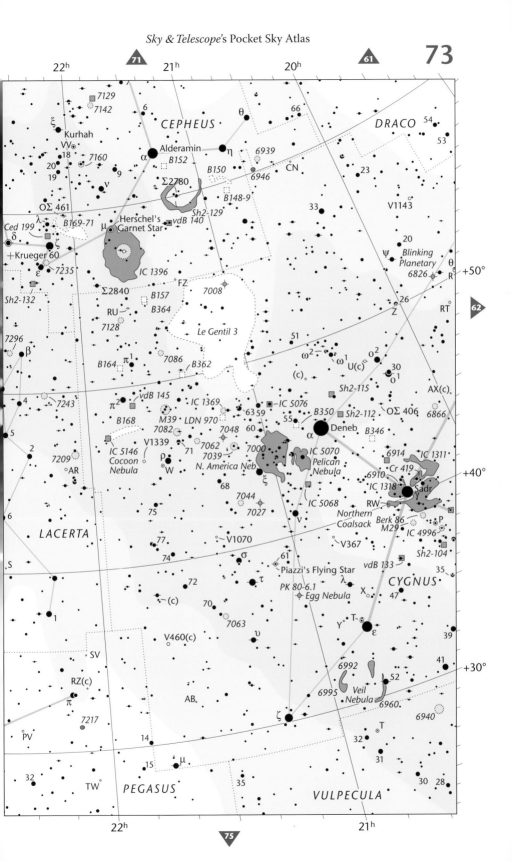

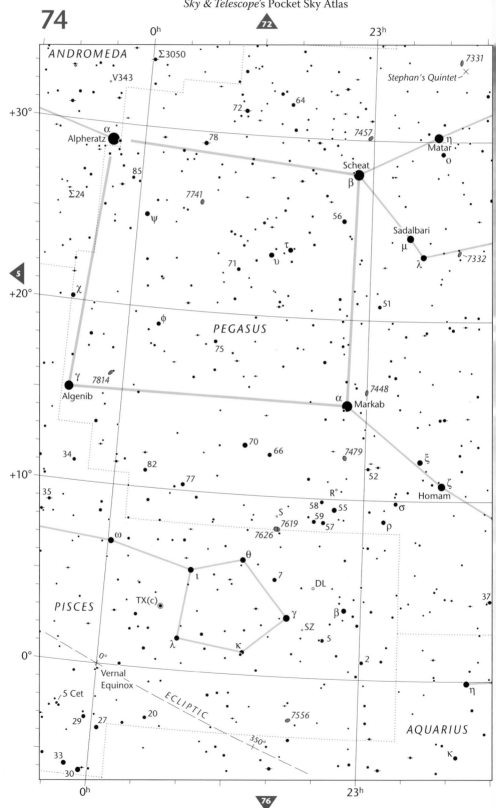

ANDROMEDA

Σ3050

7331

Stephan's Quintet

V343

64

72

78

7457

α

Alpheratz

η

Matar

ο

85

Scheat

β

Σ24

7741

56

Sadalbari

μ

ψ

7332

τ

λ

71

υ

χ

51

φ

PEGASUS

75

γ

7814

α

Markab

Algenib

7448

70

66

7479

ξ

34

82

52

ζ

77

Homam

35

R

58

55

σ

S

59

ρ

ω

7619

57

7626

θ

ι

7

DL

TX(c)

γ

β

37

PISCES

SZ

λ

κ

5

Vernal

Equinox

2

η

5 Cet

ECLIPTIC

29

27

20

7556

AQUARIUS

33

κ

30

350°

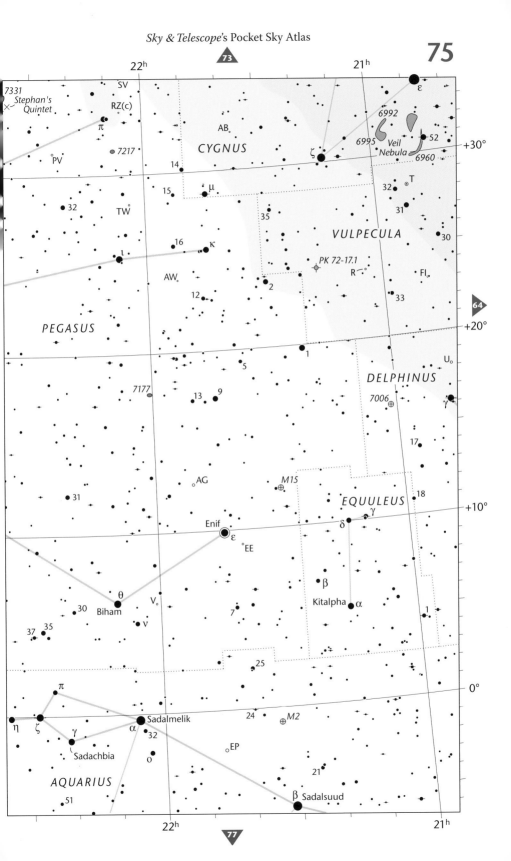

23ʰ

ω

θ

ι

7

DL

β

γ

SZ

5

35

37

π

η

ζ

TX(c)

λ

κ

2

Vernal
Equinox

0°

PISCES

ECLIPTIC

7556

AQUARIUS

κ

51

20

350°

29

27

φ

83

λ

340°

7721

χ

7606

ψ¹

ψ²

σ

33

30

3

ψ³

94

7727

7723

ω²   ω¹

97

τ

W

R

MCG-3-1-15

Ced 211

δ   Skat

6

104

106   103

2

108

66

AC

68

7

101

98

88

S

υ

7293

99

CETUS

89   ER

*Helix
Nebula*

7377

86

45

7314

ε   ζ

24

**PISCIS
AUSTRINUS**

7507

κ² κ¹

δ

SCULPTOR

Y

α

Blanco 1

ζ

Fomalhaut

Δ241

μ

γ

δ

β

S

7793

γ

0ʰ

23ʰ

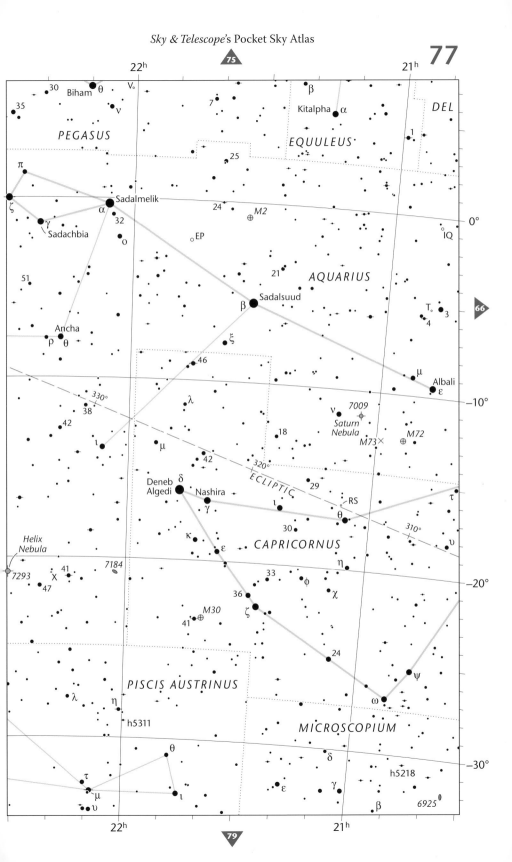

22ʰ

21ʰ

V

DEL

30   Biham   θ

35

ν

β

Kitalpha   α

1

PEGASUS

7

EQUULEUS

25

π

Sadalmelik

24

ζ

γ

α

32

Sadachbia

ο

0°

M2

EP

IQ

51

21

AQUARIUS

Ancha

β   Sadalsuud

T   3

ρ   θ

4

66

ξ

46

μ

Albali

λ

ε   −10°

330°

ν   7009

38

Saturn

M72

42

18

Nebula

M73

ι

μ

42

M30 area...

320°

ECLIPTIC

Deneb   δ

29

Algedi

Nashira

ι

RS

τ

γ

θ

Helix

κ

30

υ

Nebula

ε

CAPRICORNUS

310°

7293

41

7184

η

−20°

X

33

47

φ

36

χ

M30

ζ

41

24

ψ

PISCIS AUSTRINUS

ω

λ

η

MICROSCOPIUM

h5311

θ

δ

−30°

τ

ε   γ   h5218

μ

ι

υ

β   6925

76

0^h　　　　23^h

PISCIS AUSTRINUS　ε

CETUS

45

24

7507

κ^2　κ^1

δ

α

Fomalhaut

150

ι

Blanco 1　ζ

γ°

Δ241

δ

−30°

μ

γ

γ

7793

S

π

Lacaille 9352

η

IC 5332

IC 1459

7418

134

θ

7713　β

υ

7410

σ^2

55

7462

φ

7424

ρ

λ^2

λ^1

IC 5325

SX

ι

7590　7582

7599　7552

7412

θ

IC 5267

7531　7496

α　Ankaa

PHOENIX

IC 5328

ι

−40°

κ

θ

β

ε

τ^3

τ^1

h3390

GRUS

μ

λ^2

τ

ε

λ^1

R

Al　β

σ

ζ

κ

η

ρ

π

o

7689

DM

−50°

CI

S

γ

ζ

q^1　*ERI*

ξ

Δ5 = p Eri　α

η

TUCANA

Achernar

ν

HYI

β

η

IC 5250

2^h　　　1^h　　　0^h　　　23^h

10
80

9

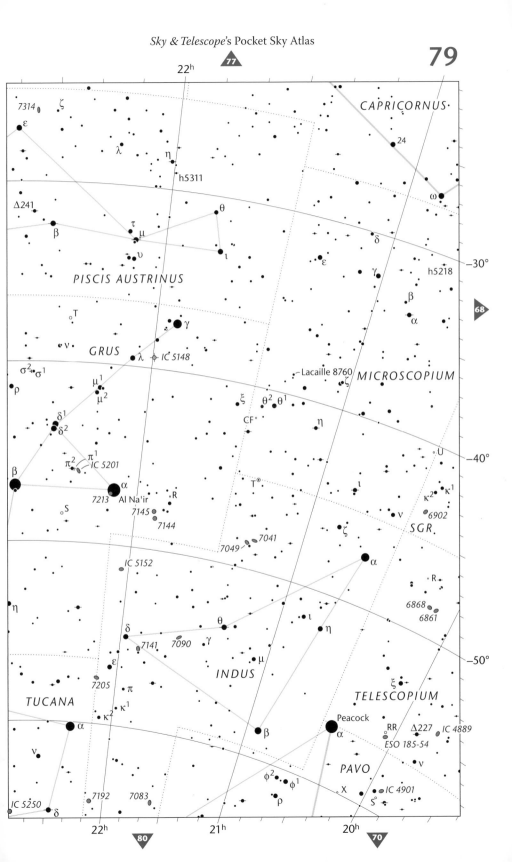

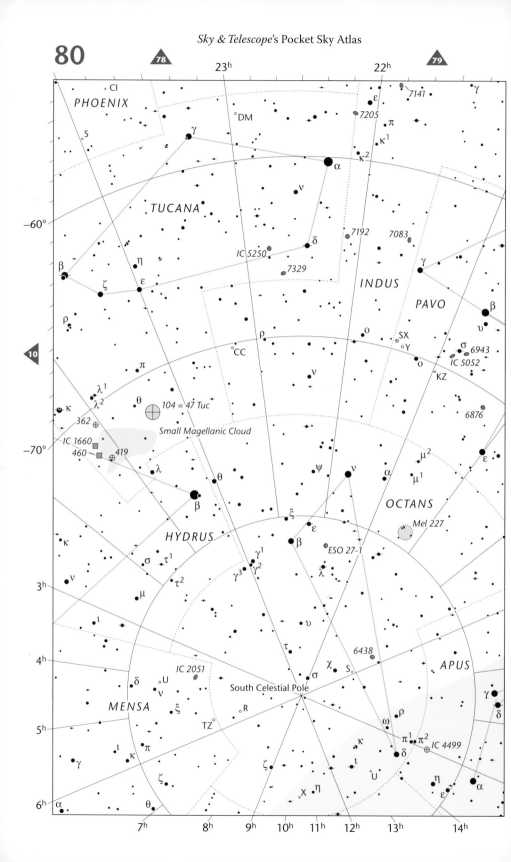

Close-up Charts

Т he following four charts show some of the most-observed regions of the sky. Compared to those in the main atlas, these charts have a larger scale and include fainter stars and more detail in nebulous areas.

Chart A covers the Pleiades and some of the nebulosity around individual stars. This is the best-known open star cluster in the heavens, but here we omit the dotted

continued on Chart D

A

3ʰ 50ᵐ	3ʰ 48ᵐ	3ʰ 46ᵐ	3ʰ 44ᵐ

Pleiades (M45)
(Close-up for Charts 13 & 15)

Sterope

Taygeta
19

Maia Nebula
1432

Maia
20

Celaeno
16

+24.5°

vdB 23   β536

Pleione
28 = BU

Atlas
27

Alcyone   η

Electra
17

+24.0°

Merope
23

IC 349

Merope Nebula
1435

TAURUS

+23.5°

Star Magnitudes										
2	3	4	5	6	7	8	9	10	11	12

3ʰ 50ᵐ	3ʰ 48ᵐ

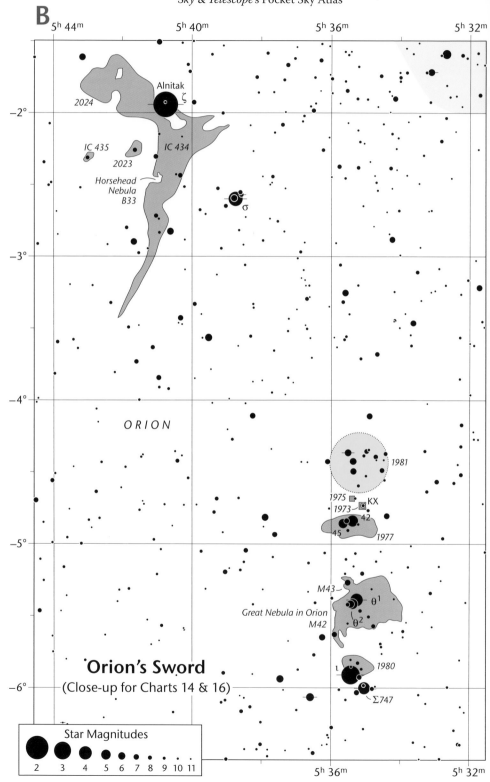

B

5ʰ 44ᵐ 5ʰ 40ᵐ 5ʰ 36ᵐ 5ʰ 32ᵐ

−2°

Alnitak
2024 ζ

IC 435 IC 434
2023
Horsehead
Nebula
B33

σ

−3°

ORION

−4°

1981

1975
1973 KX
42
45 1977

−5°

M43
θ¹
Great Nebula in Orion θ²
M42

Orion's Sword

(Close-up for Charts 14 & 16)

ι 1980

−6°

Σ747

Star Magnitudes

2 3 4 5 6 7 8 9 10 11

5ʰ 36ᵐ 5ʰ 32ᵐ

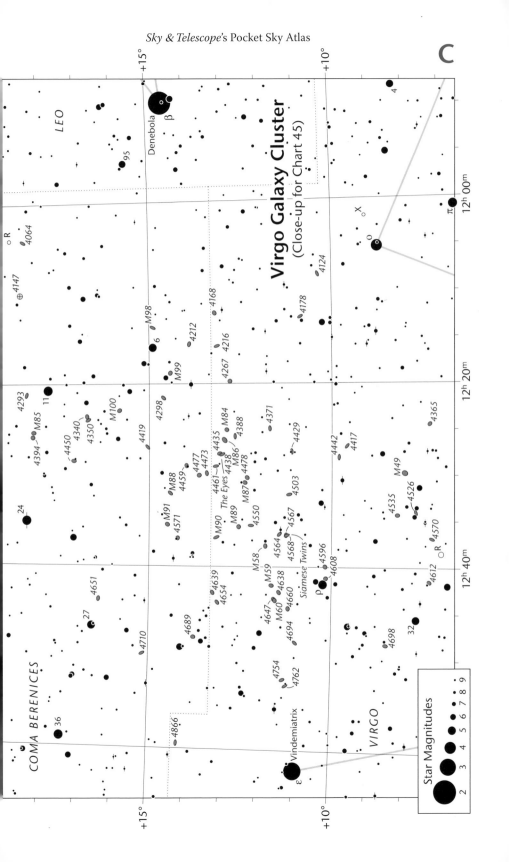

C

Virgo Galaxy Cluster
(Close-up for Chart 45)

LEO

COMA BERENICES

VIRGO

Star Magnitudes

2 3 4 5 6 7 8 9

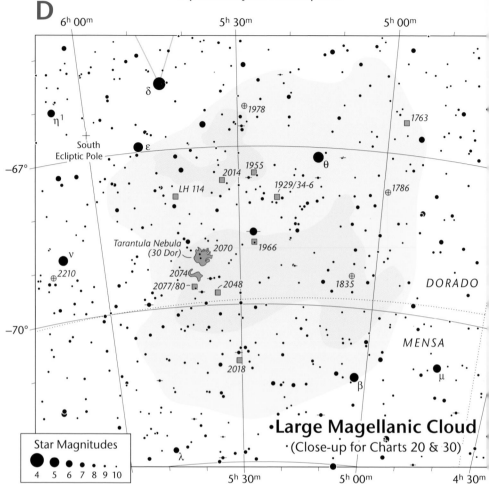

D

6ʰ 00ᵐ   5ʰ 30ᵐ   5ʰ 00ᵐ

δ

⊕1978

η¹

1763

+
South
Ecliptic Pole   ε

θ

−67°

2014   1955

LH 114   1929/34-6

⊕1786

Tarantula Nebula
(30 Dor)   2070   1966

ν

2074

2210   2077/80   2048   1835   DORADO

−70°

MENSA

2018

β

μ

•Large Magellanic Cloud
•(Close-up for Charts 20 & 30)

Star Magnitudes

λ

4   5   6   7   8   9   10   5ʰ 30ᵐ   5ʰ 00ᵐ   4ʰ 30ᵐ

continued from Chart A
yellow open-cluster symbol for clarity.

Chart B includes the easternmost bright star of Orion's Belt, Alnitak (ζ Orionis), and the softly glowing nebulosity IC 434 that makes the dark Horsehead Nebula visible in silhouette. A few degrees to the southwest is the incomparable region of Orion's Sword, containing M42, the Orion Nebula.

Visual comet hunters traditionally avoid the Virgo Galaxy Cluster, shown in Chart C, where many galaxies masquerade as comet suspects. With the help of this chart, however, the brightest of these galaxies can be identified.

Chart D, the Large Magellanic Cloud, shows only a handful of the deep-sky objects that can be seen in small telescopes within this satellite galaxy of our own Milky Way.

General Index

For stars with common names and nonstellar objects, this listing gives the number (1–80) of a main chart or the letter (A–D) of a close-up chart on which they can be found. Objects are grouped by type: stars, galaxies (and groups), open star clusters, globular clusters, bright nebulae (including supernova remnants), dark nebulae, planetary nebulae, and other. A constellation index appears on page xii.

Bold type is used to indicate a chart that shows the object well (that is, not too near an edge). In the rare cases when an object is plotted but not labeled on a given chart (due to being near an edge or in a crowded field), the label can be found on another chart of the same region.

Globular Clusters

Objects in the Caldwell Catalog

Patrick Moore, Britain's renowned popularizer of astronomy, compiled the Caldwell Catalog of 109 deep-sky objects in 1995. Some are just as spectacular as the best Messier objects; others are astrophysically intriguing.

The Go To mounts of many telescopes allow objects to be located using their C (Caldwell) numbers alone. While C numbers are not used in this atlas, these objects can be found on the charts by means of their corresponding NGC, IC, or other designation listed below.

Bold type is used to indicate any chart that shows the object well (that is, not too near an edge).

C1 (188), **1, 11, 21, 31,
41, 51, 61, 71**
C2 (40), **1, 71**
C3 (4236), **31, 41**
C4 (7023), **61, 71**
C5 (IC 342), 1, **11**
C6 (6543), **51, 61**
C7 (2403), **21**
C8 (559), **1, 2, 3**
C9, Cave Nebula (Sh2-155), **71, 72**
C10 (663), **1, 2**
C11, Bubble Nebula (7635), 3, **71, 72**
C12 (6946), **61, 62, 73**
C13 (457), **1, 3, 72**
C14, Double Cluster (869/884), **1, 2, 13**
C15, Blinking Planetary (6826), **62,** 73
C16 (7243), 62, **73**
C17 (147), **3, 72**
C18 (185), **3, 72**
C19, Cocoon Nebula (IC 5146), **62, 73**
C20, North America Nebula (7000), **62, 73**
C21 (4449), **32, 43**
C22 (7662), **3, 72**
C23 (891), **2, 13**
C24 (1275), **2, 13**
C25 (2419), **23**
C26 (4244), **32, 43**
C27, Crescent Nebula (6888), **62,** 73
C28 (752), **2**
C29 (5005), 32, **43**

C30 (7331), **72,** 74, 75
C31 (IC 405), **12, 14**
C32 (4631), **32, 43, 45**
C33, eastern Veil Nebula (6992/5), **62, 64, 73, 75**
C34, western Veil Nebula (6960), **62, 64, 73, 75**
C35 (4889), 32, 43, **45**
C36 (4559), **32, 43, 45**
C37 (6885), **62, 64**
C38 (4565), **32, 43, 45**
C39, Eskimo Nebula (2392), **25**
C40 (3626), **34**
C41, Hyades, **15**
C42 (7006), **64, 75**
C43 (7814), **5, 74**
C44 (7479), **74**
C45 (5248), **44**
C46, Hubble's Variable Nebula (2261), **25**
C47 (6934), **64**
C48 (2775), **24, 35**
C49, Rosette Nebula (2237-8/46), **25**
C50 (2244, in Rosette), **25**
C51 (IC 1613), **5, 7**
C52 (4697), **45, 47**
C53 (3115), **37**
C54 (2506), **26**
C55, Saturn Nebula (7009), **77**
C56 (246), **7**
C57, Barnard's Galaxy

(6822), **66**
C58 (2360), **27**
C59, Ghost of Jupiter (3242), **36, 37**
C60, northern galaxy (4038) of Antennae, **36, 47**
C61, southern galaxy (4039) of Antennae, **36, 47**
C62 (247), **7**
C63, Helix Nebula (7293), **76, 77**
C64 (2362), **27,** 29
C65 (253), **7, 9**
C66 (5694), **46, 48, 57, 59**
C67 (1097), **6, 8,** 17, **19**
C68 (6729), 67, **69**
C69, Bug Nebula (6302), **58**
C70 (300), **9**
C71 (2477), **28**
C72 (55), **9,** 78
C73 (1851), **18**
C74 (3132), **39**
C75 (6124), **58**
C76 (6231), **58**
C77, Centaurus A (5128), 48, **49**
C78 (6541), **58, 69**
C79 (3201), **39**
C80, ω Centauri (5139), 48, **49,** 59
C81 (6352), **58, 69**
C82 (6193), **58**
C83 (4945), 38, **49**

C84 (5286), **48, 59**
C85 (IC 2391), **28, 39**
C86 (6397), **58, 69**
C87 (1261), **8, 19, 20**
C88 (5823), **48, 59,** 60
C89 (6087), 48, **58, 60**
C90 (2867), **28, 39, 40**
C91 (3532), **38, 40, 49**
C92, η Carinae Nebula (3372), **38, 40,** 49
C93 (6752), **69, 70**
C94, Jewel Box (4755), **38, 49, 50**
C95 (6025), 48, 58, 59, **60**
C96 (2516), **28, 30, 39**
C97 (3766), **38, 40, 49**
C98 (4609, in Coalsack), 38, **49, 50**
C99, Coalsack, 38, **49, 50**
C100 (IC 2944), **38, 40,** 49, 50
C101 (6744), **70**
C102, Southern Pleiades (IC 2602), **38, 40**
C103, Tarantula Nebula (30 Doradus, 2070), **20, 30, D**
C104 (362), **10,** 20, **80**
C105 (4833), 40, **50**
C106, 47 Tucanae (104), **10, 80**
C107 (6101), 50, **60,** 70
C108 (4372), **40, 50**
C109 (3195), 10, 20, **30, 40, 50, 60**

Objects in the Messier Catalog

Many of the best deep-sky showpieces were first tabulated by French comet hunter Charles Messier (1730–1817) or his collaborator, Pierre Méchain (1744–1805). In this atlas the Messier (M) number takes precedence over any other designa- tion. Listed below, in parentheses, is the corresponding NGC or IC designation for most of these objects.

Bold type is used to indicate any chart that shows the object well (not too near an edge).

M1, Crab Nebula (1952), **14**

M2 (7089), **75, 77**

M3 (5272), 43, **44**

M4 (6121), **56, 58**

M5 (5904), **55, 57**

M6, Butterfly Cluster (6405), **56, 58, 67, 69**

M7 (6475), **58, 67, 69**

M8, Lagoon Nebula (6523), **67, 69**

M9 (6333), **56**

M10 (6254), **54, 56**

M11, Wild Duck Cluster (6705), 65, **67**

M12 (6218), **54, 56**

M13, Hercules Cluster (6205), **52**

M14 (6402), **54, 56,** 67

M15 (7078), **75**

M16, Eagle Nebula (6611), **67**

M17, Omega Nebula (6618), **67**

M18 (6613), **67**

M19 (6273), **56, 58**

M20, Trifid Nebula (6514), **67, 69**

M21 (6531), **67, 69**

M22 (6656), **67, 69**

M23 (6494), **67**

M24 (star cloud), **67**

M25 (IC 4725), **67**

M26 (6694), **67**

M27, Dumbbell Nebula (6853), **62, 64**

M28 (6626), **67, 69**

M29 (6913), **62, 73**

M30 (7099), **77**

M31, Andromeda Galaxy (224), **3,** 72

M32 (221), **3,** 72

M33 (598), 2, **3, 4, 5**

M34 (1039), **2, 13**

M35 (2168), **12, 14, 23, 25**

M36 (1960), **12, 14**

M37 (2099), **12, 14, 23, 25**

M38 (1912), **12,** 14

M39 (7092), **62, 73**

M40 (a double star), **32, 41, 43**

M41 (2287), **27**

M42, Orion Nebula (1976), **16, B**

M43 (1982), **16, B**

M44, Beehive Cluster (2632), **24,** 35

M45, Pleiades, 13, **15, A**

M46 (2437), 26, **27**

M47 (2422), 26, **27**

M48 (2548), **26**

M49 (4472), **45, C**

M50 (2323), **27**

M51, Whirlpool Galaxy (5194-5), **32, 43**

M52 (7654), 3, **71, 72**

M53 (5024), **45**

M54 (6715), **67, 69**

M55 (6809), **66, 68**

M56 (6779), **63, 65**

M57, Ring Nebula (6720), **63,** 65

M58 (4579), **45, C**

M59 (4621), **45, C**

M60 (4649), **45, C**

M61 (4303), **45,** 47

M62 (6266), **56, 58**

M63, Sunflower Galaxy (5055), **32, 43**

M64, Black-Eye Galaxy (4826), **45**

M65 (3623), **34**

M66 (3627), **34**

M67 (2682), **24, 35**

M68 (4590), **47, 49**

M69 (6637), **67, 69**

M70 (6681), **67, 69**

M71 (6838), **64**

M72 (6981), **66, 77**

M73 (clump of four stars), 66, **77**

M74 (628), 4, **5**

M75 (6864), **66**

M76, Little Dumbbell (650-1), **2, 13**

M77 (1068), **4, 6**

M78 (2068), **14, 16**

M79 (1904), **16,** 18

M80 (6093), **56**

M81 (3031), 21, **31**

M82 (3034), 21, **31**

M83 (5236), 46, 47, **48**

M84 (4374), **45, C**

M85 (4382), **45, C**

M86 (4406), **45, C**

M87 (4486), **45, C**

M88 (4501), **45, C**

M89 (4552), **45, C**

M90 (4569), **45, C**

M91 (4548), **45, C**

M92 (6341), **52, 63**

M93 (2447), **26,** 28

M94 (4736), **32, 43**

M95 (3351), **34**

M96 (3368), **34**

M97, Owl Nebula (3587), **32, 43**

M98 (4192), 34, **45, C**

M99 (4254), **45, C**

M100 (4321), **45, C**

M101 (5457), 32, **42, 53**

M103 (581), **1,** 2, 3, 72

M104, Sombrero Galaxy (4594), **47**

M105 (3379), **34**

M106 (4258), **32, 43**

M107 (6171), **56**

M108 (3556), 31, 32, 33, **43**

M109 (3992), **32, 43**

M110 (205), **3,** 72